EDUCATION AND THE KNOWLEDGE SOCIETY

T0192004

IFIP – The International Federation for Information Processing

IFIP was founded in 1960 under the auspices of UNESCO, following the First World Computer Congress held in Paris the previous year. An umbrella organization for societies working in information processing, IFIP's aim is two-fold: to support information processing within its member countries and to encourage technology transfer to developing nations. As its mission statement clearly states,

> *IFIP's mission is to be the leading, truly international, apolitical organization which encourages and assists in the development, exploitation and application of information technology for the benefit of all people.*

IFIP is a non-profitmaking organization, run almost solely by 2500 volunteers. It operates through a number of technical committees, which organize events and publications. IFIP's events range from an international congress to local seminars, but the most important are:

- The IFIP World Computer Congress, held every second year;
- Open conferences;
- Working conferences.

The flagship event is the IFIP World Computer Congress, at which both invited and contributed papers are presented. Contributed papers are rigorously refereed and the rejection rate is high.

As with the Congress, participation in the open conferences is open to all and papers may be invited or submitted. Again, submitted papers are stringently refereed.

The working conferences are structured differently. They are usually run by a working group and attendance is small and by invitation only. Their purpose is to create an atmosphere conducive to innovation and development. Refereeing is less rigorous and papers are subjected to extensive group discussion.

Publications arising from IFIP events vary. The papers presented at the IFIP World Computer Congress and at open conferences are published as conference proceedings, while the results of the working conferences are often published as collections of selected and edited papers.

Any national society whose primary activity is in information may apply to become a full member of IFIP, although full membership is restricted to one society per country. Full members are entitled to vote at the annual General Assembly, National societies preferring a less committed involvement may apply for associate or corresponding membership. Associate members enjoy the same benefits as full members, but without voting rights. Corresponding members are not represented in IFIP bodies. Affiliated membership is open to non-national societies, and individual and honorary membership schemes are also offered.

EDUCATION AND THE KNOWLEDGE SOCIETY
Information Technology supporting human development

Edited by

Tom J. van Weert
Hogeschool van Utrecht, The Netherlands

Kluwer Academic Publishers
Boston/Dordrecht/London

Tom J. van Weert
Hogescoo van Utrecht
The Netherlands

Library of Congress Cataloging-in-Publication Data

A C.I.P. Catalogue record for this book is available from the Library of Congress.

EDUCATION AND THE KNOWLEDGE SOCIETY: Information Technology Supporting Human Development / Edited by Tom J. van Weert
   p.cm. (The International Federation for Information Processing)

ISBN 978-1-4419-5431-2   Printed on acid-free paper.
e-ISBN 978-0-387-23120-4

Printed in the United States of America.

9 8 7 6 5 4 3 2 1
springeronline.com

Contents

Contributing Authors

Jacques Berleur (Belgium)
Klaus Brunnstein (Germany)
Roger Carlsen (USA)
Bernard Cornu (France)
Annie Corsini-Karagouni (Switzerland)
Niki Davis (UK-USA)
Kenneth Deer (Canada)
Steven Dijkxhoorn (Netherlands)
Anita Erkelens (Netherlands)
Antoine Geissbuhler (Switzerland)
Mohan R. Gurubatham (Malaysia)
Anneke E.N. Hacquebard (Netherlands)
Ann-Kristin Håkansson (Sweden)
Colin Harrison (USA-Switzerland)
André Hurst (Switzerland)
Willis King (USA)
Jean-Marie Leclerc (Switzerland)
René Longet (Switzerland)
Ousmane Ly (Mali)
Sithabile Magwizi (Zimbabwe)
André-Yves Portnoff (France)
Mikko J. Ruohonen (Finland)
Wesley Shrum (USA)
J. Barrie Thomson (United Kingdom)
Jarmo Viteli (Finland)
Deryn Watson (United Kingdom)

Tom J. van Weert (Netherlands)
David Wood (United Kingdom)

Preface

Engineering the Knowledge Society (EKS) - Event of the World Summit on the Information Society (WSIS)

This book is the result of a joint event of the World Federation of Engineering Organisations (WFEO) and the International Federation for Information Processing (IFIP) held during the World Summit on the Information Society (WSIS) in Geneva, Switzerland, December 11 - 12, 2003. The organisation was in the hands of Mr. Raymond Morel of the Swiss Academy of Engineering Sciences (SATW).

Information Technology (or Information and Communication Technology) cannot be seen as a separate entity. Its application should support human development and this application has to be engineered. Education plays a central role in the engineering of Information and Communication Technology (ICT) for human support. The conference addressed the following aspects: Lifelong Learning and education, e-inclusion, ethics and social impact, engineering profession, developing e-society, economy and e-Society. The contributions in this World Summit event reflected an active stance towards human development supported by ICT. A Round Table session provided concrete proposals for action.

The International Programme Committee of this WSIS event was formed by:
- Jean-Claude Badoux (WFEO, SATW)
- Leszek J. Bialy (WFEO)
- Pierre-André Bobillier (IFIP, SFI, SI, SISR)
- Fulvio Caccia (SATW)

- Louis-Joseph Fleury (SATW)
- Raymond Morel (IFIP, SATW, SFI, SI, SISR)
- Andreas Schweizer (SATW)
- Tom van Weert (IFIP, SATW)

The International Program Committee of Engineering the Knowledge Society (EKS) wishes to thank the following partners who made this event of the World Summit on the Information Society possible:
- WFEO (World Federation of Engineering Organisations);
- SATW (The Swiss Academy of Engineering Sciences);
- The Hasler Stiftung;
- SUN Microsystems;
- SWITCH (Swiss Academic and Research Network or Swiss education & Research Network);
- SVI-FSI (Swiss federation of Informatics Societies);
- SI (Swiss Informatics Society);
- ICT Switzerland;
- SISR (Société d'Informatique de Suisse Romande).

Papers

This book has been produced from papers by invited authors from Belgium, Canada, Finland, France, Germany, Malaysia, Mali, The Netherlands, Switzerland, United Kingdom and the USA. In addition the book contains project proposals for taking action. Also included are the UNESCO - IFIP World Computer Congress 2002 Youth Declaration, UNESCO - IFIP Vilnius Declaration, the World Federation of Engineering Organisations (WFEO) Carthage Declaration on the Digital Divide and the European Academy of Sciences and Arts (EASA) Declaration on security of Wireless Lans.

Foreword

This book presents a cross-section of educational implications of the Knowledge Society (or Information Society), an emerging society in which Information and Communication Technology plays an important role. In the words of Klaus Brunnstein, the IFIP president: "Renowned experts address essential aspects such as the engineering of ICT-applications for the Knowledge Society, as well as the vulnerability of such applications. Also the concept of a networked economy and an example of telemedicine in Mali are presented. Sustainability is addressed, thus linking Rio to Geneva. More general, very basic aspects of knowledge-based systems are addressed: roles and preservation of information, education and lifelong learning, the development of ICT-skills and professionalism, social engineering, as well as questions related to ethics in the information society.

With such richness in detail an essential basis is built for understanding the relation between the concept of the 'Information Society' and the educational aspects of the 'Knowledge Society' as discussed in the 2003 General Conference of UNESCO with special reference to the needs of developing knowledge societies."

The papers are structured according to the following themes:
- Lifelong Learning and education;
- e-Inclusion;
- Ethics and social impact;
- The role of engineering;
- Developing e-society;
- Economy and e-Society;
- Project proposals.

Editor

Tom J. van Weert holds the chair in ICT and Higher Education of the Hogeschool van Utrecht, University of Professional Education and Applied Science, Utrecht, The Netherlands. Earlier he was managing director of Cetis, centre of expertise for educational innovation and ICT, of the same university. Before that he was director of the School of Informatics (Computing Science) at the University of Nijmegen, The Netherlands. Tom has studied applied mathematics and computing science. He started his working career in teacher education and software engineering. He has been chair of the International Federation for Information Processing (IFIP) Working Groups on Secondary Education and Higher Education. He currently is vice-chair of IFIP Technical Committee 3 (TC3) on Education. He is also member of the TC3 Taskforce on Lifelong Learning.

A comprehensive synthesis of research
Information and Communication Technology in education

Niki Davis & Roger Carlsen
Iowa State University Center of Technology in Learning and Teaching, Iowa State University Ames, Iowa 50011-3192, USA. Also Institute of Education, University of London, UK
nedavis@iastate.edu

Department of Educational Leadership, Graduate Educational Technology, School of Education and Human Services, Wright State University, Dayton, Ohio 45435, USA
roger.carlsen@wright.edu

Abstract: The World Summit on the Information Society (WSIS) is working to connect educational and community institutions to those with low resources, who are suffering challenges that can be hard for others to grasp. The aim is to build capacity using Information and Communication Technology (ICT). In this paper a comprehensive synthesis of research into ICT in education is given with the aim to inform the WSIS planning and action by drawing together reviews of policy and research, with an emphasis on formal education in schools and universities. The synthesis provides a picture of a complex situation by considering four reasons to invest in ICT in education: economic competitiveness, ICT to increase educational attainment, ICT to increase access to education, and ICT as a catalyst for educational renewal. It clarifies the complex challenges for such reengineering, which will require sustained interdisciplinary and intercultural collaboration to support the diversity of our emerging knowledge societies. This synthesis is a work in progress. The authors invite feedback and additional references to reviews of research, especially to research findings for the many underrepresented populations of the world.

Key words: access, curriculum, economic competitiveness, educational attainment, educational renewal, ICT, primary, research synthesis, secondary, university

INTRODUCTION

The World Summit on the Information Society (WSIS), in Geneva in 2004 and in Tunis in 2005, is working to extend the principles enshrined in the Charter of the United Nations and the Universal Declaration of Human Rights to include Information and Communication Technologies (ICT) because of the many key roles of ICT in our current Information Society. The *Draft Plan of Action* that was published for WSIS in November 2003 included objectives for the connection of educational and community institutions with ICT and "to adapt all primary and secondary curricula to meet the challenges of the Information Society, taking into account national circumstances" (page 2). Action lines include "Capacity Building", in terms of ICT-literacy and the use of ICT to "eradicate illiteracies" and "to empower local communities, especially those in rural and underserved areas"(page 5). The WSIS website provides illustrative stories of successes in digital information at:
   http://www.itu.int/osg/wsis-themes/ict_stories/DigitalEducation.html.
The WSIS event "Engineering the Knowledge Society", where this paper was presented, took place alongside national leaders' debates and emphasised interdisciplinary views and the imperative to support diversity.

The comprehensive synthesis of research into ICT in education, presented in this paper, aims to inform the WSIS planning and action by drawing together reviews of policy, with an emphasis on formal education in schools and universities. Education in the subject of Informatics or Computing Science is excluded; this is discussed elsewhere, for example in (Mulder & van Weert 1998) and (Anderson & van Weert 2003).

Our editors' request to limit references has increased our selectivity to use around 20 reviews and influential studies that include international research from over 400 researchers in over 200 countries on 4 continents. The research is mainly derived from North America, Europe and Australia, which is another aspect of the lack of social justice and equity that pervades society worldwide. Figure 1 provides a very brief overview of the evidence in relation to ICT in primary, secondary and university education. The discussion is structured around four reasons for investing in ICT in education:
1. Economic competitiveness;
2. ICT to increase educational attainment;
3. ICT to increase access to education;
4. ICT as a catalyst for educational renewal.

IMPERATIVE FOR ECONOMIC COMPETITIVENESS

The economic impetrative to invest in ICT in education was reviewed for high income countries by the Partnership for 21st Century Skills (2003) who considered the impact of technology on the job market, the flow of information and resources in a global marketplace, and the impact of digital technologies on daily life. They conclude that 21st century skills are about "use of knowledge and skills – by thinking critically" (Partnership for 21st Century Skills 2003; p. 9). In Europe "thinking critically" plus sensitivity to culture is also proposed for a European I-curriculum (Ulicsak & Owen 2003).

1. ICT as a imperative for economic competitiveness
   - 21st century literacies should incorporate critical and cultural views of ICT;
   - Operational knowledge of ICT is better in vocational education;
   - Curriculum development has system-wide implications.

2. ICT to increase educational attainment
   - ICT's general support is unreliable;
   - Many topics can be supported by ICT, e.g. calculators for algebra;
   - ICT can be used to support creative assessment;
   - ICT support requires sophisticated pedagogic understanding.

3. ICT to increase access to education
   - ICT can enable access to teachers, learners and/or content;
   - ICT can enable access for those with special educational needs;
   - Use of ICT can reduce dependence and increase autonomy;
   - Indigenous people may adapt ICT to their cultural needs.

4. Educational renewal – ICT as a catalyst
   - Innovation by early adopters does not spread easily;
   - Change is slow and ICT blends into current practice to 'stretch the mould';
   - ICT augments central control first;
   - Teacher education is required to extend and adapt pedagogic knowledge.

Figure 1. ICT in education structured by four imperatives

The earlier review by the Educational Multimedia Task Force 1995-2001 (Belisle et al. 2001) reported that many learners, teachers and trainers were becoming computer literate and able to use common ICT-tools without help. Their review noted that "literacy for the digital era requires new core competencies," with "a fundamental shift from the traditional teaching paradigm to the self supported learning paradigm" that leads to a change in the design of schools.

This synthesis questions the teaching of ICT operational skills in primary, secondary and university education. Somekh (Brown & Davis, 2004) reports the growth of UK children's conceptual understanding of computer networks and their relative maturity compared to adults. Downes reported that despite relatively good access to ICT in schools, children in Australia, UK and the USA rarely used computers in school in a satisfying way due to tensions that relate to time, control and the nature of learning (Brown & Davis, 2004). Downes links this to the "affordance of the computer as a "playable tool" that facilitates the blending of play, practice and performance," which is an approach rarely encouraged in formal education.

Therefore, critical and cultural aspects of ICT appear to be the important aspects to incorporate into formal curricula, but ICT operational skills appear to be best left to informal and vocational education. The UNESCO planning guide for ICT in Teacher Education (UNESCO 2002) provides the most comprehensive guide for such curriculum planning, with a rationale and framework that could be applied to ICT curriculum planning at many levels. Its holistic nature facilities interpretation into target cultures in line with recent research on the nature of learning, which clarifies the integrative and contextual nature of learning. UNESCO's holistic framework has four competencies (Content & pedagogy, Technical issues, Social issues, and Collaboration & networking) that are embedded in four overarching themes (Leadership & vision, Lifelong learning, Context & culture, and Planning & the management of change).

ICT TO INCREASE EDUCATIONAL ATTAINMENT

Many research studies have demonstrated a positive impact of ICT on education, which may have led to a general belief that ICT can improve attainment in education. However, the positive impact only in rare cases appears to extend to increased attainment for wider populations. For example Shachter's (1999) review of over 700 empirical research studies reviewed evidence of the positive educational gains from ICT in a variety of forms including, integrated learning systems, and a few negative forms, including the use of drill and practice software, with interactions between teacher technology training and more positive school climate. Large scale studies of integrated learning systems, see (Underwood & Brown, 1997), and recently the more general impact of ICT (BECTa, 2003a) in the United Kingdom showed small statistically significant gains, but the evidence is not compelling, given the investment required to align curriculum, assessment, teacher education and access to ICT in school and beyond. In addition, these reviews rarely discuss the opportunity costs that have been highlighted by the Alliance for Childhood who, in their provocative review Fool's Gold, draw attention to alterative ways that money could be spent, lost educational opportunities, and health and developmental risks; see: (Kirkpatrick & Cuban, 1998).

Similar optimistic beliefs relating to ICT in universities were dispelled by a study in the United Kingdom, led by the first author (Boucher, Davis et al. 1997). The review of the research literature, case studies and economic analysis of ICT showed widespread application, but limited economic effectiveness. Although economic measures proved impossible to isolate, positive educational impact was visible where:

– Routine or mechanical skills play an important part (e.g. graphing calculators);
– Knowledge that can be precisely specified for identified users;
– Well defined professional base.

There are many topics within the curriculum where ICT has shown benefits when accompanied by changes in the curriculum and teacher education. Perhaps the best illustration is the graphing calculator, which is widely used to teach algebra. This ICT- tool can be used to increase higher level thinking skills when pedagogy and curricula are adjusted sympathetically; see for example (Valdez et al. 2000). There is also evidence that ICT can assist in assessment, which becomes more demanding when ICT is used to enhance learning (EPPI 2003). However such changes to curricula are accompanied by debate on the loss of basic skills, such as mental arithmetic in this instance. According to Bull et al. (Johnston & Maddux 2003), the graphing calculator, which was specially designed for education, is a model for more affordable ubiquitous computing in the future. Handheld computers may also be more adaptable than larger computers to low income contexts.

ICT FOR ACCESS TO EDUCATION

In the context of WSIS increasing access to education may be more effective than topic specific educational gains discussed in the last section. There is widespread evidence that ICT can increase access to content, teachers and learners and that ICT is already used for this purpose worldwide. For learners with special educational needs ICT can enable access to education and improve their integration into society.

For over a decade ICT has been used effectively by adult learners for access to education moving through at least three generations of distance learning. Radio and TV are widely used in low income countries with rapid expansion of web-based learning in high income countries. IITE (2000) reviews policies, pedagogy and professional development in detail and notes that distance education "is a complex activity requiring a policy and regulatory framework, appropriate organization and pedagogic structures, funding and professional staff, coordination and quality assurance." The establishment of "virtual universities" has provided new economic models that also increase access. Research of the use of the Internet in the USA (Pew 2002) shows extensive use of the Internet by students for schoolwork (94% of youth ages 12-17, mainly from outside school). Also over 50% of adults have used the Internet for school or job training and for work-related research. The report notes that students "are coming to school with different expectations, different skills and access to different resources" and they "insist that policy makers take the digital divide seriously." The recent expansion of virtual schooling, using web-based learning environments across the USA, has also been researched during its development to provide evidence that aligns with the IITE report: successful use of ICT to access education is related to individual student's self-motivation and self-study skills, as well as experience with technology and a good attitude towards the subject matter.

An increasingly wide set of solutions continues to be developed to respond to special educational needs. Individual case studies provide compelling evidence of beneficial impacts; see (BECTa 2003b) and Jeffs et al. in (Johnston & Maddux 2003). Jeffs et al. provide a 20 year retrospective analysis of technological advancements in special education with benefits for those who have learning disabilities, mental retardation, deaf/hard of hearing, and for the gifted and talented. They note the complexity of the infusion of technology and the need for educators to be aware of their own pedagogical beliefs. Successful ICT applications include head operated switches, Braille output and software that can be adjusted to individual needs. A wide range of software has been developed, and the many challenges in adapting it to local curricula and languages, other than English, is likely to continue to be supported by dedicated volunteers. It is notable that teachers have used software packages to individualize materials for specific learners to facilitate integration of disabled learners into everyday life, including appropriate voice annotations and assessment, with accompanying gains in self esteem and improved function. In addition, the use of communication technologies to network teachers dedicated to specific needs has enabled them to expand their knowledge. There are projects implementing many of these applications worldwide, some highlighted in WSIS case studies, but adoption beyond the first world may remain low due to economic factors.

PROMOTION OF CHANGE – ICT AS A CATALYST

We now turn to the final imperative, which is the use of ICT as a catalyst to promote educational renewal. There is widespread evidence in many sectors that once ICT is introduced it continues to catalyse change; the nature of ICT is to pervade a system. Ellsworth's (2000) survey of change models can facilitate understanding of the processes from a number of perspectives. Stages of ICT adoption have been identified and simplified for both organisations and educators, and it is possible that they could be extended to regions and learners. For example Valdez et al. (2000) describe evolving uses and expectations for organisations: "print automation, expansion of learning opportunities and data-driven decision making." "Early adopters" are the first to innovate with ICT. An international study of classrooms innovating with ICT in 26 countries found a very wide variety of applications and pedagogy. The common thread was an emerging paradigm of lifelong learning (Pelgrum & Anderson 2001). A complementary study in Hong Kong noted that these innovative practices did not spread; see: Law in (Brown & Davis 2003). This evidence of the lack of diffusion of innovation with ICT is supported by data from the USA that teachers are not using ICT in ways recommended by research (McMillan et al. 2003).

Research informed by theory permits gathering of more sophisticated evidence. "Hypothesis-driven multiple case studies" researched by Wood (2003) suggest that, in the view of those European leaders interviewed, primary and secondary education are unlikely to radically change with ICT, because neither radical change nor collapse of the educational system was envisaged. Instead these leaders recognised that, although ICT currently augments centralised control, and this would continue, control must also be relaxed to engender a research and development role for teachers due to fundamental uncertainty about ICT in education. The major concerns of

these leaders are illuminating, because these illustrate the complexity of change with ICT in education:
- Current assessment of students is poorly aligned with recent research on learning and pedagogy for ICT;
- Required pedagogic knowledge has been grossly underestimated;
- The general public and society remain to be convinced on ICT in education;
- All educational professionals are implicated;
- Content and services providers are changing with economies of scale.

These leaders also see a range of ethical, legal and validity issues emerging, including copyright issues and quality and health issues of networks of people linked by ICT. A scenario approach to research of change with ICT in universities produced complementary findings that ICT is widespread, occurring as a blend, and that change is slow, not radical (Collis & Wende 2003). Universities that accept the challenge of lifelong learning change the most.

Countries and regions that have used ICT to catalyse renewal of education include Chile, rural regions of Hungary and tele-learning communities across Canada (respectively described by Hepp et al., Turcsanyi-Szabo, and Laferriere et al., in: (Brown & Davis 2003). All these projects required significant leadership at local and central points, plus particular care to include minority cultures. Resta et al., in: (Brown & Davis 2004) illustrate the very different world views that are held by many indigenous nations, and they provide guidance with model protocols and strategies illustrated with an example from the Anishinable nation that is indigenous in North America. ICT has also acted as catalyst in education with leadership from multinationals. For example, Cisco Networking Academies in over 150 countries deliver a suite "of web-based and instructor led vocational courses in ICT skills" Design in the USA has been supported by over US$200 million donated by Cisco as part of its corporate social responsibility; see Selinger in: (Brown & Davis 2004). The content is mediated by local teachers with classroom and lab-based instruction. Cisco's Academy initiative has helped regions and nations to go online for the first time, including the training of educational leaders in Africa. Some institutions have used the Academy to prepare the way for their own web-based courses. This is one of many illustrations where multinational interests in ICT can coincide with those of the local population but, in a review of free and open software, Rajani et al. (2003) caution as to the foreign culture and values that may be unwittingly imported into education through "closed" software. We extend this caution to content, pedagogy and curricula.

WSIS AND ENGINEERING OUR KNOWLEDGE SOCIETIES

This comprehensive synthesis of research into ICT in education challenges objectives stated in the draft plan of action for WSIS. Research on the addition of ICT to education has not provided clear evidence of economic or educational gains, although it is clear that ICT has catalysed change in education alongside changes in other sectors of society in the first world, where ICT has become prevalent. Research evidence is also mainly from the first world, with limited surveys and research on ICT initiatives in countries beyond Europe, North America and Australia. This is an additional aspect of social inequity that is known as the "digital divide." The direct application of research findings from the first world to lower income contexts is cautioned.

Although this comprehensive synthesis supports the expansion of literacy in the twenty-first century to include critical and cultural aspects of ICT, it finds that operational skills of ICT may be taught more effectively in vocational education. Research into the use of ICT to enhance educational attainment is complex, suggesting the need for complex and systemic engineering with extensive access to ICT for students and their teachers within and beyond education. However, there is clear evidence of the successful application of ICT to increase access to education, such as in rural areas. ICT to increase access for special educational needs is also indicated. However, these applications of ICT to enhance access also require systematic and systemic engineering, involving many partners, including teachers, who would do well to gain an understanding of their complementary motivations to invest in ICT, so that they may work in harmony.

Other chapters of this book provide complementary perspectives from relevant disciplines including engineers and sociologists who have experience in low income regions. Taken together we suggest that the WSIS plans for capacity building with ICT through education will require sustained interdisciplinary collaboration to engineer diversity within our emerging knowledge societies. As teacher educators and researchers we hope to play our part; see for example the project proposal by Davis in this volume.

This comprehensive synthesis of ICT in education is also a work that we will continue.. WSIS in Geneva has provided the first staging post and others are planned, including activities of the Working Group on research of the International Federation of Information Processing, Technical Committee 3 on Education. We look forward to receive feedback on this synthesis and references to reviews of research that may be added.

BIOGRAPHY

Niki Davis is Director of Iowa State University Center of Technology in Learning where she leads the graduate program in Curriculum and Instructional Technology that is well known for its emphasis in teacher education. She also holds a Chair in ICT in Education at the Institute of Education, University of London, where she is a member of the London Knowledge Lab. Before this she held a chair in Educational Telematics in the University of Exeter in the UK where she set up the Telematics Centre.

Niki has researched information technologies extensively, particularly in teacher education and in flexible and distance learning. She is currently the President of the international Society of Information Technology in Teacher Education and Chair of the International Federation of Information Processing Technical Committee 3 on Education's Working Group on Research. She is also an invited expert of UNESCO on ICT teacher education.

Roger Carlsen is the program advisor and faculty leader for the graduate educational technology program at Wright State University. He recently served as a member of the National Educational Technology Standards writing team. Roger is a program editor for the Society for Information Technology & Teacher Education and a member of International Federation of Information Processing Technical Committee for Education's Working Groups 3.6   Research and 3.3   Distance Education. At Wright State University, Roger is chair-elect for the College of Education and Human Services Educational Technology Committee. He is the faculty Senate's representative to the University's Portal Committee and also serves on the University Center for International Education Committee to internationalize the curriculum. In addition to a broad and deep knowledge of educational technology, Roger's specialty areas are program evaluation, the use of forums in education, and statistics. His most recent work involves using technology to educate emerging cultures.

REFERENCES

Anderson, J. & T. J. van Weert (eds.) (2002) *Information and Communication Technology in education, a curriculum for schools and programme for teacher development.* UNESCO, Division of Higher Education, Paris.

Belisle, C., A. Rawlings & C. van Seventer (2001) *The Educational Multimedia Task Force 1995-2001. Integrated research effort on multimedia in education and training.* European Commission, Luxembourg.

BECTa (2003a*) ImpaCT2 Learning at home and school: Case studies.* ICT in schools research and evaluation series, No. 8. British Educational Communications and Technology Agency: Coventry, United Kingdom. [http://www.becta.org/research/impact2]

BECTa (2003b) *What the research says about ICT supporting special educational needs (SEN) and inclusion.* ICT in schools research and evaluation series, No. 12. British Educational Communications and Technology Agency: Coventry, United Kingdom. [http://www.becta.org/research/]

Boucher, A., N. E. Davis, P. Dillon, P. Hobbs & P. Tearle (1997) *Information Technology Assisted Teaching and Learning in Higher Education.* HEFCE Research Series. Higher Education Funding Council for England, Bristol.

Brown, A. & N. E. Davis (eds.) (2004) *Digital technology, communities and education.* World Yearbook in Education 2004. Routledge, London.

Collis, B. & M. van Wende (eds.) (2002) *Models of technology and change in higher education. An international comparative survey on the current and future use of ICT in higher education.* EDEN [http://www.eden.bme.hu/contents/dissemination/bulletin.html]

Ellsworth, J. (2000) *Surviving change: A survey of educational change models.* ERIC: Syracuse, NY.

EPPI (2003) *A systematic review of the impact on students and teachers of the use of ICT for assessment of creative and critical thinking skills.* Evidence for Policy and Practice Information and Coordination Centre, Institute of Education, University of London, United Kingdom. [http://eppi.ioe.ac.uk/]

IITE (2000) *Analytical survey of distance education for the Information Society: policies, pedagogy and professional development.* UNESCO Institute for Information Technologies in Education, Moscow. [http://www.iite.ru]

Johnson, D.L. & C. D. Maddux (eds.) (2003) *Technology in education. A twenty year retrospective.* The Hayworth Press, Binghampton, NY.

Kirkpatrick H. & L. Cuban (1998) Computers make kids smarter, right? *Technos Quarterly, 7, 2.* [http://www.technos.net/tq_07/2cuban.htm]

McMillan Culp, K., M. Honey & E. Mandinach (2003) *A retrospective on twenty years of education policy.* US Department of Education, Washington DC, USA. [http://www.nationaledtechplan.org/participate/20years.pdf]

Mulder, F. & T.van Weert (eds.) (1998) *Informatics in Higher Education.* Chapman & Hall, London.

Partnership for 21st Century Skills (2003) *Learning for the 21st century.* Author, Washington DC. [http://www.21stcenturyskills.org]

Pelgrum, W.J. & R. E. Anderson (2001) *ICT and the emerging paradigm for lifelong learning. An IEA educational assessment of infrastructure, goals, and practices in twenty-six countries.* Second edition. IEA, Amsterdam. [http://www.iea.nl/Home/IEA/Publicaions/SITES_book.pdf]

Pew (2003) *Pew Internet and American life.* The Pew Research Center. [http://www.pewinternet.org/reports/]

Rajani N., J. Rekola & T. Mielonen (2003) *Free as in Education. Significance of the free/libre and open source software for developing countries.* Ministry of Foreign Affairs of Finland Department of Public Policy, Helsinki, Finland.

Schachter J. (1999) Does technology improve students' learning and achievement? How, when, and under what conditions? *Journal of Educational Computing Research, 20.* Milken Exchange on Educational Technology. [http://www.mff.org]

Ulicsak M. & M. Owen (2003) *I-Curriculum.* Personal communication.

Underwood, J. & J. Brown (1997) *Integrated Learning Systems: Potential into Practice.* Heinemann, London.

UNESCO (2002) *ICT in teacher education. A planning guide.* UNESCO: Paris, France. [http://unesdoc.unesco.org/images/0012/001295129533e.pdf]

Valdez G., M. McNabb, M. Foertsch, M. Anderson, M. Hawkes. & L. Raack (2000) *Computer-based technology and learning: Evolving uses and expectations.* North Central Regional Educational Laboratory, Illinois. [http://www.ncrel.org/tplan/cbtl/toc.htm]

Wood, D. (2003) *Think again: Insight, insight and foresight on ICT in schools.* Draft report for EUN's ERNIST project. November 2003, personal communication.

LIFELONG LEARNING IN THE KNOWLEDGE SOCIETY
Implications for education

Tom van Weert

Hogeschool van Utrecht, University for Professional Education, The Netherlands; Cetis, P. O. Box 85029, 3508 AA Utrecht, The Netherlands; Tel: + 31 30 258 6296, Fax: +31 30 258 6292

t.vweert@cetis.hvu.nl; http://www.cetis.hvu.nl

Abstract: A Knowledge Society is developing in which Information and Communication Technology is both a catalyst and a necessity. Knowledge is an invaluable asset in this ICT-integrated society, both *tacit knowledge* in the heads and hands of the workers and *explicit knowledge*. *Human capital* is becoming more and more important. Innovation is driving force in knowledge intensive economies. Therefore application and creation of new knowledge are normal part of the work of modern professionals: lifelong working implies *lifelong learning*. Students are the professionals of tomorrow and need to develop the competences of the knowledge worker. This implies a change in educational paradigm and educational transformation to *new education*. For developing countries and transition economies this offers threats and opportunities.

Key words: education, human capital, innovation, knowledge, knowledge worker, learning organisation, pedagogy, real-life learning, student, transformation

ICT-INTEGRATED KNOWLEDGE SOCIETY

A Knowledge Society is developing in which Information and Communication Technology is both a catalyst and a necessity. Knowledge is an invaluable asset in this ICT-integrated society where production, services, consumption and trade are rapidly changing. To keep up with developments (knowledge) workers need to adapt continuously and acquire new competences: working and learning melt together. At the work place new knowledge is created to keep up with developments: both tacit knowledge in the heads and hands of the workers and explicit knowledge (codified, operational knowledge). The concept of knowledge is changing from scientific, theoretical knowledge ('old knowledge') to more operational knowledge ('new knowledge'). Human capital is becoming more and more important and workers become more and more responsible for all dimensions of their work.

KNOWLEDGE ECONOMY AND ICT

"The concepts of *'knowledge economy'* and *'knowledge worker'* are based on the view that information and knowledge are at the centre of economic growth and development. The ability to produce and use information effectively is thus a vital source for skills of many individuals" (2000b). "Technological change and innovation drive the development of the knowledge-based economy through their effects on production methods, consumption patterns and the structure of economies. Both are closely related in recent growth performance. Some changes in innovation processes could not have occurred without ICTs and conversely, some of the impact of ICTs might not have been felt in the absence of changes in the innovation system." (OECD 2000a).

WORKING IS LEARNING

"A knowledge-based economy relies primarily on the use of ideas rather than physical abilities and on the application of technology rather than the transformation of raw materials or the exploitation of cheap labour. Knowledge is being developed and applied in new ways. Product cycles are shorter and the need for innovation greater. Trade is increasing worldwide, increasing competitive demands on producers." (World Bank 2002b, p.ix). In the knowledge economy, change is so rapid that workers constantly need to acquire new skills. Firms need workers who are willing and able to update their skills throughout their lifetimes. (ibid)

To keep up with developments in a knowledge-intensive economy (knowledge) workers need to adapt continuously to new developments: they are in a process of Lifelong Learning. "In the old economy, the basic competences of the industrial worker, bricklayer, or bus driver were relatively stable. True, you might have applied these competencies to different situations, such as different construction sites, but the learning component of your labour was small. In the new economy, the learning component of work becomes huge. .... Think about your own work. Work and learning overlap for a massive component of the workforce." (Tapscott 1996, p. 198).

KNOWLEDGE CREATION AT THE WORK PLACE

"In the knowledge economy the term knowledge was used originally to denote scientific knowledge However, partly under the influence of Information and Communication Technology the concept of knowledge is broadening: knowledge, wherever it is stored, becomes available. Knowledge in the heads of or hands of workers can be codified; tacit knowledge can be a commercially valuable asset" (WRR 2002; p. 22; author's translation). Mass distribution of ICT and the Internet seem to contribute towards the development of new knowledge and new attitudes towards knowledge. The concept of knowledge has been extended from purely theoretical knowledge ('old knowledge') to knowledge that is also more practice-oriented ('new knowledge'). (WRR 2002)

HUMAN CAPITAL

Long term developments lead to fundamental changes in economic activities and put more weight on unique human qualities such as knowledge creation. Robotic type of work is taken over by automates. "To adapt and maintain competitiveness in response to changing consumer preferences and technological change, companies need appropriate organisational structures, a skilled workforce and able management. These changes are having a significant impact on the structure of employment and on the type of labour required. The most obvious manifestation of this is the rising human capital levels of the populations and workforces in OECD countries, as measured by the educational attainment and as implied by an increased demand for more highly-educated and highly-skilled workers." (OECD 2001; p. 102) Human capital is becoming more and more important and allows workers more freedom in giving form to their work commitments. Supported by Information and Communication Technology they become more and more responsible for all dimensions of their work. This contributes to the 'wholeness' of working life. "More and more people give meaning to their lives in paid professional work. The reason for this is the changing character of work. By and by an 'enterprising' society of dynamic professionals is developing." (Beek 1998).

THE LEARNING ORGANISATION

"These changes also have affected the way in which organisations interact in the economy, with networking, co-operation and the fluid flow of knowledge within and across national borders gaining in importance." (OECD 2001; p. 100/101) "To keep up with demands and competition innovative businesses and organisations have to create new operational knowledge in their domain: how to do better and how to offer new products and/or services. In a learning organisation work is organised in non-traditional ways and professionals work in a different way. A shift can be observed from organisational structures suited for efficient, standard, large-scale throughput (Tayloristic, old economy) to structures that facilitate flexible, custom-tailored, small scale, high quality production or servicing (networked, new economy). These new organisational structures aim to satisfy a personal, demand-driven market and are reflected in organisational concepts such as "Just In Time". The new structures are geared towards teamwork, flexibility and quality. Information and Communication Technology (ICT) is omnipresent and empowers the individual to act as expert in many areas. ICT also offers flexibility in time and place in support of teamwork." (Weert 2002).

WORKING AND LEARNING MELT TOGETHER

Innovation is driving force in knowledge intensive economies. Therefore the economical focus is on knowledge work: new application of existing knowledge and knowledge creation. Knowledge application and knowledge creation are normal part of the work of modern professionals. The modern professional is a knowledge worker for whom lifelong working implies *lifelong learning*.

Knowledge workers

Knowledge work is organised in non-traditional ways. Aim is to satisfy demand-driven market and the organisation of work is geared towards teamwork, flexibility and quality. Information and Communication Technology (ICT) is omnipresent and empowers the individual to act as expert in many areas. ICI also offers flexibility in time and place in support of teamwork. Work is result oriented and the professionals are accountable on results: team and organisation form a meritocracy.

The new professionals give meaning to their lives through their work. They continually engage in new challenges and learn on the job. They therefore need other skills than in the 'old fashioned' Tayloristic economy. "Social-communicative and social-normative skills and competences (soft skills or people skills) are needed to be able to function adequately in teams and cooperate with colleagues: communication skills, empathy, team player skills. Self-direction and autonomy require initiative, pro-activity, flexibility and risk taking of professionals." (WRR 2002; p. 148). Another of these new qualifications is the capability to deal with a professional environment characterized by fast change. For the modern professional lifelong working is identical with Lifelong Learning; the modern professional is a learning professional." (Weert 2004).

Table 1. Traditional versus Lifelong learning

Traditional learning model	Lifelong learning
The teacher is the source of knowledge	Educators are guides to sources of knowledge
Learners receive knowledge from the teacher	People learn by doing
Learners work by themselves	People learn in groups and from each other
Tests are given to prevent progress until students have completely mastered a set of skills and to ration access to further learning	Assessment is used to guide learning strategies and identify pathways for future learning
All learners do the same thing	Educators develop individualized learning plans
Teachers receive initial training plus ad hoc in-service training	Educators are lifelong learners. Initial training and on-going professional development are linked
"Good" learners are identified and permitted to continue their education	People have access to learning opportunities over a lifetime

STUDENTS AS THE KNOWLEDGE WORKERS OF TOMORROW

"The first question to be answered is: "What is learning". Here we assume that learning is the use and the creation of new operational knowledge (Go & van Weert 2004) that steers our actions. Learning is a social activity in which interactions with the environment (human and non-human) play an important role. As the IFIP Focus Group Report (2004) states: "Traditional education methods are ill suited to providing people with the skills they need to be successful in a knowledge economy. The traditional learning model differs from lifelong learning methods in important ways. Table 1., emphasizing these differences, is taken from a World Bank Lifelong Learning report (World Bank 2002b, p.xi).

There is need for a paradigm change from 'acquisition view' to 'constructivism view' as presented by Duffy (Table 2.)."

Table 2. Contrasting views of learning (Duffy 2001)

	Constructivism view	Acquisition view
Learning is	Organic	Cumulative
	Continual reorganisation	Discovery (finding what is known)
	Invention	
Knowledge is	A *construction*	An *acquisition*
Coach-apprentice relation is	Mutual respect for views Ability to converse	Transfer of coach's expertise
Assessment is	Ability to use knowledge	Mastery of content

NEW EDUCATION

Real-life learning

Lifelong Learning takes place in a real life setting where new knowledge has to be created and applied. In this real life context learning is aimed at developing operational knowledge to perform better as a professional or to become a starting professional. Working as a professional may be characterised by the following three phases (Simons 2001):

1. *relate*: working with knowledge, learning-on-the-job and making explicit the implicit results of learning;
2. *create*: extending knowledge by, for example, carrying out research, explicit learning;

3. *donate*: putting into practice, presenting, promoting one's own knowledge, contribute to the profession.

Professional development is not part of everyday schoolwork in class, so it needs an extra effort. Universities should work in co-operation with business, industry and other organisations in the field to develop and implement programmes which are intended to give students optimum preparation for the reality and dynamics of professional practice (Go & van Weert 2004). These new programmes aim to provide learning environments which enable students to develop into starting professionals: they develop their competence and professional expertise in learning environments of varying complexity. The introduction of new competence-based programmes means adopting a new methodology and this assumes that the tutor also develops competencies and professional expertise in relation to the programme (Witteman 2001).

New pedagogy

A new pedagogy is needed in education for learning in authentic, real-life situations. Herrington, Oliver, and Reeves (2002) have defined ten design principles for developing and evaluating authentic activity-based learning environments. Authentic activities must:
1. Have real-world relevance.
2. Be ill defined, requiring students to define the tasks and sub-tasks needed to complete the activity.
3. Comprise complex tasks to be investigated by students over a sustained period of time.
4. Provide the opportunity for students to examine the task from different perspectives, using a variety of resources.
5. Provide the opportunity to collaborate.
6. Provide the opportunity to reflect and involve students' beliefs and values.
7. Be integrated and applied across different subject areas and extend beyond domain-specific outcomes.
8. Be seamlessly integrated with assessment.
9. Yield polished products valuable in their own right rather than as preparation for something else.
10. Allow competing solutions and diversity of outcomes.

Herrington et al. (2004) state: "In direct contrast to the academic approach, practical problems tend to be characterized by: the key roles of problem recognition and definition, the ill-defined nature of the problem, substantial information seeking, multiple correct solutions, multiple methods of obtaining solutions, the availability of relevant prior experience, and often highly motivating and emotionally involving contingencies (Sternberg et al., 1993, p. 206). Key differences between the school-based approach and real-life approach are summarised in Table 3 (Lebow & Wager 1994).

Table 3. Real-life versus in-school problem solving (Lebow & Wager 1994)

Real-life	In-school
Involves ill-formulated problems and ill-structured conditions	Involves text-book problems and well-structured conditions
Problems are embedded in specific and meaningful context	Problems are largely abstract and de-contextualised
Problems have depth, complexity and duration	Problems lack depth, complexity and duration
Involves cooperative relations and shared consequences	Involves competitive relations and individual assessment
Problems are perceived as real and worth solving	Problems typically seem artificial with low relevance for students

Institutional transformation

Educational institutions will be forced to creatively rethink and renew institutional organisation. As Büetner et al. (2004) state: "ICT becomes an integral though invisible part of the daily personal productivity and professional practice. The focus of the curriculum is becoming learner-centred and integrates subject areas in real-world applications. For example, students may work with community leaders to solve local problems by accessing, analysing, reporting, and presenting information with ITC tools. Learner's access to technology is broad and unrestricted. They take more responsibility for their own learning and assessment. The institution has become a centre of learning for the business community."

When transforming, educational institutions will have to develop (Büetner et al. 2004):

- Vision;
- Philosophy of learning and pedagogy;
- Development plans and policies;
- Facilities and resources;
- Understanding the curriculum;
- Professional development of institution staff;
- Communities;
- Assessment.

DEVELOPING COUNTRIES AND TRANSITION ECONOMIES

The Knowledge Society impacts on our global environment, entailing opportunities and threats for developing countries and transition economies. See Table 4 (World Bank 2002A; p.8).

Table 4. Opportunities and threats stemming from changes in the global environment

Change factor	Opportunities	Threats
Growing role of knowledge	Possibility of leapfrogging in selected areas of economic growth Resolution of social problems (food security, health, water supply, energy, environment)	Increasing knowledge gap among nations
ICT revolution	Easier access to knowledge and information	Growing digital divide among and within nations
Global labour market	Easier access to expertise, skills, and knowledge embedded in professionals	Growing brain drain and loss of advanced human capital
Political and social change Spread of democracy Violence, corruption, and crime HIV/AIDS	Positive environment for reform	Growing brain drain and political instability Loss of human resources

"Tertiary education institutions have a critical role in supporting knowledge-driven economic growth strategies and the construction of democratic, socially cohesive societies. Tertiary education assists the improvement of the institutional regime through the training of competent and responsible professionals needed for sound macroeconomic and public sector management. ..... To successfully fulfil their educational, research, and informational functions in the 21st century, tertiary education institutions need to be able to respond effectively to changing education and training needs, adapt to a rapidly shifting tertiary education landscape, and adopt more flexible modes of organization and operation." (World Bank 2002a, p.23)

"Developing countries and transition economies risk being further marginalized in a competitive global knowledge economy because their education and training systems are not equipping learners with the skills they need. To respond to the problem, policymakers need to make crucial changes. They need to replace the information-based, teacher-directed, directive-based rote learning provided within a formal education system with a new type of learning that emphasizes creating, applying, analyzing, and synthesizing knowledge and engaging in collaborative learning throughout the lifespan." (World Bank 2002b; p.ix).

ACKNOWLEDGEMENT

In this paper results of the IFIP Focus Group Report on Lifelong learning in the digital age (IFIP Lifelong Learning Focus Group 2004) have been used.

BIOGRAPHY

Tom J. van Weert holds the chair in ICT and Higher Education of the Hogeschool van Utrecht, University of Professional Education and Applied Science, Utrecht, The Netherlands. Earlier he was managing director of Cetis, centre of expertise for educational innovation and ICT, of the same university. Before that he was director of the School of Informatics (Computing Science) at the University of Nijmegen, The Netherlands. Tom has studied applied mathematics and computing science. He started his working career in teacher education and software engineering. He has been chair of the International Federation for Information Processing (IFIP) Working Groups on Secondary Education and Higher Education. He currently is vice-chair of IFIP Technical Committee 3 (TC3) on Education. He is also member of the TC3 Taskforce on Lifelong Learning.

REFERENCES

Beek, K. van (1998) *De ondernemende samenleving. Een verkenning van maatschappelijke veradering en implicaties voor beleid.* Wetenschappelijke Raad voor het Regeringsbeleid, Voorstudies en achtergronden V104, Sdu Uitgevers, Den Haag.

Duffy, Th. M. & Ch. Orrill (2001) *Constructivism,* in Kovalchic A. & K. Dawson (Eds.), Educational Technology, An Encyclopedia. Santa Barbara, CA: ABC-CLIO.

Go, F. & T. J. van Weert (2004) *Regional knowledge networks for lifelong learning,* in: Weert, T. J. van & Kendall, M. (Eds.) (2004) Lifelong Learning in the Digital Age. Kluwer Academic Publishers, Boston/Dordrech/London.

Herrington, Jan, Thomas C. Reeves, Ron Oliver & Younghee Woo (2004) Designing Authentic Activities in Web-Based Courses. Journal of Computing in Higher Education, Fall 2004, Vol. 16(1), 3-29.

IFIP Lifelong Learning Focus Group (2004) *Focus Group Report on Lifelong learning in the digital age,* in: Weert, T. J. van & Kendall, M. (Eds.) (2004) Learning in the Digital Age. Kluwer Academic Publishers, Boston/Dordrecht/London.

OECD (2000a) *A new economy? The changing role of innovation and information technology in growth.* Organisation for Economic Co-operation and Development, Paris.

OECD (2000b) *Knowledge management in the learning society.* Organisation for Economic Co-operation and Development, Paris.

OECD (2001) *Educational Policy Analysis 2001.* Centre for Educational Research and Innovation Organisation for Economic Co-operation and Development (OECD), Paris, http://www.oecd.org.

Reeves, T.C., J. Herrington & R. Oliver (2002). Authentic activities and online learning. In A. Goody, J. Herrington & M. Northcote (Eds) *Quality conversations: Research and Development in Higher Education,* Volume 25 (pp. 562-567). Jamison, ACT: HERDSA. [verified 6 Feb 2003] http://www.ecu.edu.au/conferences/herdsa/papers/ref/pdf/Reeves.pdf

Simons, P.R.J., & M. Ruijters (2001) *Learning professionals: towards an integrated model,* paper presented at the biannual conference of the European Association for Research on learning and Instruction, Fribourg (Switzerland).

Tapscott, D. (1996) *Digital Economy, Promise and peril in the age of networked intelligence.* McGraw-Hill.

Weert, T. J. van (2002) *Lifelong learning in Virtual Learning Organisations, Designing virtual learning environments*, in: Passey, D. & M. Kendall (Eds.) (2002) TelE-LEARNING, The Challenge for the Third Millennium, Kluwer Academic Publishers, Boston/Dordrecht/London, p. 135 -142.

Weert, T. J. van (2004) *New higher education for lifelong learning*, in: Weert, T. J. van & Kendall, M. (Eds.) (2004) Learning in the Digital Age. Kluwer Academic Publishers, Boston/Dordrech/London. In preparation.

Witteman, H. (2001) Oude didactiek past niet in nieuwe onderwijsvisies, in: *Thema, Tijdschrift voor Hoger Onderwijs en Management*, april 2001, nr.1, 17-25.

World Bank (2002a) *Constructing Knowledge Societies: New Challenges for Tertiary Education.* A World Bank report. Washington, D.C. http://www1.worldbank.org/education/tertiary/lifelong.asp

World Bank (2002b) *Lifelong Learning in the Global Knowledge Economy: Challenges for Developing Countries.* World Bank, Washington D.C. http://www1.worldbank.org/education/pdf/Lifelong%20Learning_GKE.pdf

Wetenschappelijke Raad voor het Regeringsbeleid (WRR) (2002) *Van oude en nieuwe kennis, De gevolgen van ICT voor het kennisbeleid*, Sdu Uitgevers, Den Haag. http://www.wrr.nl/HTML-NL/BasisPU-NL.html

Collective intelligence and capacity building
In a networked society

Bernard Cornu

Professor, Director of La Villa Media, 46 avenue Felix Viallet, 38000 Grenoble, France
bernard.cornu@grenoble.iufm.fr

Abstract: The Information Society has new and specific characteristics. Information is digitalised, and therefore interactive, processable, transportable, accessible in new ways. Information is changing more quickly than before, and we see new types of information appear. The Information Society is a networked society, a society where collective capabilities are more and more needed, in addition to individual ones. Capacity building in the Information Society is not just an extension of the usual capacity building. New competences, new capabilities are necessary. These are not only technological competences, but more profound competences linked with new concepts. Not only competences linked to new knowledge, but competences linked to new ways to access knowledge. In particular three aspects of capacity building in the Information Society will be analysed: Accessing and processing knowledge in a networked world; Collective intelligence and collective capacity, and New citizenship in the Information Society. New capabilities cannot be acquired through the old ways of education: capacity building needs new contents and methods.

Key words: capacity building, collective intelligence, competences, education, ethics, information society, knowledge, network

THE INFORMATION SOCIETY

The changes in information and communication technologies have been huge during the last two decades. It is not only a matter of technology, computers and computer science, but a profound change has taken place in society. Information is digitalized, in huge quantities of "0's" and "1's". The processing and transportation of digitalized information involve similar processes and tools. This has lead to a merging of information and communication technologies which we call "ICT", Information and Communication Technologies. Digitalised information has many new characteristics: it is fast moving (and much less stable than it was before), it is interactive (one can act on it very easily, in a "to and fro" way, one can change information), it takes new forms, such as multimedia (combining different forms of media: texts, images, sounds, movies, etc.) and hypermedia (linking media through hyperlinks). These new characteristics have profound consequences:

- Information cannot simply be acquired, stored, passed on and taught as before. Digitalised information is now accessible in new ways, stored in new conditions and in very big quantities; distribution of information can be done in different ways and at very wide scales.
- Information has become an economic good, which has its own value, which can be produced, sold, bought and exchanged.

This is why one speaks of the "Information Society": a society in which information is one of the major goods, a society that is widely based on information, a society in which economical power can be attained through information.

Changes have of course also affected education. Information is one of the major ingredients of education, but education cannot be satisfied with just information: education has mainly to deal with knowledge.

THE KNOWLEDGE SOCIETY

It would be a mistake to confuse information with knowledge. Knowledge has to do with human beings and cannot be reduced to a set of "0's" and "1's", as information is. Knowledge has an "institutionalised" dimension: it is recognised as knowledge by a community of people and is related to a corpus of knowledge with its specific uses and applications. In the information society knowledge takes a central place. It has most of the characteristics of information in terms of processing and of transportation, but knowledge cannot simply be transmitted and taught as before as it has become an economic good. In this sense one can say that new "knowledge societies" have appeared. In a simple "information society" it would be easy to circulate and disseminate information, but it would be a society without any innovation. As education deals with the transmission of knowledge, not only of information, it has to be considered in knowledge societies.

In knowledge societies not only traditional knowledge takes a new form, but knowledge is transformed. In each discipline main concepts and processes are changed. But in addition there are other kinds of knowledge, more complex, more transverse, which address the needs of society and of education of the future; see for instance the "Seven complex lessons in education for the future" by Morin (2001). We are forced to rethink what knowledge is. Access to knowledge is also profoundly changed. It is not only accessible in books and libraries, or in the teachers' head, but also accessible from a huge variety of sources, varying in certainty and accuracy, at any place at any time. In conclusion: knowledge societies create new knowledge and new ways of storing and accessing knowledge.

A NETWORKED SOCIETY

A major characteristic of "ICT societies" and knowledge societies is that these are networked. We were used to pyramidal or hierarchical structures in the way our systems are organised, in the way we access information in a book through the library directory or the table of contents, in the way we communicate with people. Information and communication technologies lead to networked structures, that for example in the Internet are particularly clear. A network is a set of nodes, linked by edges. Nodes can be of many types (often creating some confusion): information, knowledge, people, institution, etc. Edges link the nodes and can usually be followed by just a "click". New links can appear continuously, enriching the network and making it more complex. It is typical for a network that there are many different ways to go from one node to another (while in a pyramidal structure, there may be only one way). If I am not satisfied by one way, I can choose another. Networks disturb hierarchies, particularly in schools and in education. In a network there may be sub-networks, introducing new kinds of hierarchies. Communication is not only "one to one" or "one to all", it becomes "all to all": a networked communication.

The networked structure of society affects all domains of social life, having positive and negative consequences. Economy is affected, social life, leisure and politics are affected. And education is affected. Educational systems in a networked society cannot stay as they were; they must take into account the new communication facilities, the networked structure, the new ways of dealing with space and time. Teaching and learning in a networked society is different and uses new tools, new resources, new pedagogies. The teaching profession in a networked society has new dimensions and has to fulfil new expectations from society; a new kind of competence is needed.

COLLECTIVE INTELLIGENCE AND COLLABORATIVE WORK

Knowledge societies and networked societies need new and enhanced forms of intelligence, in particular collective intelligence. We are used to individual intelligence. Our educational systems are designed to develop individual intelligence and individual capabilities of each student. But development of communication and networks creates a need for collective intelligence. Just think of ants: as an individual an ant is a very simple animal, with almost no intelligence. But collectively ants can find the shortest way from one point to another, can carry heavy loads, can build bridges, can regulate the temperature in the ant-hill. This is not done in a hierarchical way, but through collective intelligence which uses indirect communication through the environment by the means of pheromones. The collective intelligence of ants is not the sum of their individual intelligences; it stems from their communication ability and their relationship to the environment.

Similarly information and communication technologies and networks allow human beings to develop collective intelligence that is much more than the sum of individual intelligence. Virtual communities are examples of such collective intelligence. The semantic web is a future means to express collective intelligence of humanity: networked humanity. Exchange and cooperation through networks and ICT contribute to the development our collective intelligence, and therefore our capability for learning. Collective intelligence makes possible new actions, new developments, new competences.

However, collective intelligence is not something new. Mankind has been using collective intelligence for a long time (to build pyramids or cathedrals, to design systems, to discover nature and to develop science and technology, etc.). But because of ICTs and networks collective intelligence turns into a major dimension in knowledge societies. Collaborative work in knowledge societies is not just another tool, but is fundamentally linked to the networked society and to collective intelligence. Education has to take this into account: develop collective intelligence of students, in the classroom and in other kinds of communities, real or virtual; develop the capacity for collaborative work, use collaborative work as a mean to develop collective intelligence.

CAPACITY BUILDING

Knowledge societies need new kinds of competences and reinforcement of some existing ones: competences for collaborating, for communicating, for being part of a virtual community, for being part of a group with collective intelligence. Education therefore is facing a new challenge. Education has to incorporate the collective dimension in learning and teaching, to practice collaborative work and to prepare students to collaborative work in a "collective intelligence" way, to develop communication skills.

Networks and the networked society need new competences for taking part in a network, working in a network, contributing to a network, communicating in a network, accessing information and knowledge in a network, collaborate with others in a network. Again, this is a challenge for education in knowledge societies. Networked environments should be introduced in the classroom, and the classroom should be incorporated in networks. Much more, the concept of a classroom or even a school will have to evolve, using the network and collective possibilities. There is no reason for the classroom to be at a single place where pupils are together with a teacher, all at the same time. Education will have to function in other kinds of spaces and times (physical and virtual spaces, synchronous and asynchronous times...). Functioning in new spaces and times is a major competence, for which we must prepare the students.

Accessing information and knowledge is also an essential competence: finding the way to such knowledge or information, sorting through huge quantities of information, evaluating the accuracy and the reliability of such information, storing, processing, sharing information and knowledge, using and applying information and knowledge, etc. It is essential that our educational systems takes such competences in account.

The famous "four pillars" of education: "learning to know, learning to do, learning to live together, learning to be", take a new dimension in knowledge societies, in networked societies. They provide an agenda for capacity building in knowledge societies.

CITIZENSHIP IN KNOWLEDGE SOCIETIES

Education not only has the task to transfer knowledge and make it internalised by students, but also has to prepare students as future citizens, to transfer values of society and humanity. Citizenship addresses ethical issues that must be taken into account in education. But citizenship in knowledge societies includes new dimensions, mainly due to networks and communication and to the collective dimension.

Citizenship is a matter of rights and duties. In knowledge societies, networks raise new questions about rights and duties, since communication is easier, and since information is accessible in new forms and through new ways. The respect for each person's dignity, the respect for privacy and social equity take new forms in networked societies. This should be an important issue for education. Citizenship also includes the participation of citizens in the social life. Networks enhance the possibilities for such participation, and decision-makers should take this into account.

With respect to education, ethical questions are even more prominent in the networked society: how can knowledge societies attain the goals of "Education for all", in terms of access and quality? How can knowledge societies cope with the issue of knowledge and education as merchandise? Knowledge is a public good, education is a public service, and this implies that accurate policies are necessary for education. Does education prepare consumers or citizens? How can education cope with globalisation? How can education contribute to reduction of the digital divide, between countries, but also within each country and even within each school or classroom?

Such issues must be addressed by education. There is a need for new educational policies for knowledge societies, integrating new forms of knowledge, new forms of access to knowledge, integrating the networked form of society.

Knowledge societies are for human beings, not for technology. Humanism may improve and increase in knowledge societies, if we are able to use accurately all the new available technologies, tools and resources.

Equity is one of the major issues in knowledge societies. We must keep aiming at "knowledge societies for all". This is a goal for "education for all" in knowledge societies.

CONCLUSION

Computers and computer science have created information and communication technologies with as a result a society based on technology and information. But the human dimension forces us to aim at a knowledge society, not only an information society. Knowledge societies have new characteristics, and education in knowledge societies is more complex, but more and more essential. Knowledge societies reinforce the expectations towards education, and the need for education.

Education and capacity building in knowledge societies is not only a matter of content and curricula; it is mainly a matter of methods, it needs major changes in the way education is performed. Education must address new issues such as processing information, processing, acquiring and transmitting knowledge, being able to act in a networked society, developing collective intelligence and collaborative work, opening up to new spaces and times, building new kinds of capacities, preparing citizens of the knowledge society. Such goals need profound changes in the methods of education.

But of course, the purpose is not to change everything and reject what was done before. It is not a revolution, but rather an evolution that we need. Addressing fully the new possibilities, the new tools and resources of the knowledge societies, does not mean rejecting fundamental knowledge and the traditional basis of education. We have to find the right balance in the tension between tradition and modernity, so that the knowledge society is really a society of humanism, a society for all, a society in which the well being of persons, the fundamental values, and the fundamental rights and duties of human beings are kept central. The knowledge society is a human society, not a technological society.

BIOGRAPHY

Prof. Bernard Cornu is the director of La Villa Media (the European Residence for Educational Multimedia), Grenoble, France, and a professor at the IUFM of Grenoble, France (University Institute for Teacher Education). He is a professor (applied Mathematics) at IUFM (Institut Universitaire de Formation des Maitres - University Institute for Teacher Education) of Grenoble, France of which he has been for ten years (1990-2000) the director. Until 1994, Bernard Cornu was the chairman of the 29 IUFMs in France. He was (2000-2002) advisor for Teacher Education at the French Ministry of Education.

Bernard Cornu is also a mathematician at Grenoble University. He studied the influence of computers and informatics on mathematics and its teaching, and also worked in didactics of mathematics. He has been the director of the Institute of Research on Mathematics Teaching (IREM) of Grenoble, and then the head at the in-service teacher training office for the Academy of Grenoble. His scientific specialism now is the integration of Information and Communication Technologies into education, and its influence on the teaching profession and on educational policies.

As a member of IFIP (International Federation for Information Processing), he has been (1995-2000) the chairman of Working Group 3.1 ("Informatics Education at the Secondary Education Level"), and he is now the secretary of the IFIP TC3 (Technical Committee for Education).

Bernard Cornu is a member of the French National Commission for UNESCO, and the vice-chair of the Education Committee. He is also the vice-chair of the Governing Board of IITE, the UNESCO Institute for Information Technologies in education, located in Moscow. He has been (1998-2002) the President of the French Commission for Mathematics Education.

REFERENCES

Morin, Edgar (2001) *Seven complex lessons in education for the future.* UNESCO Publishing, Paris.

Binde, Jerome & Jean-Joseph Goux (2003) *"0 et 1, briques du futur".* Le Monde, October 26, 2003, Paris.

*E*Tampere
Social engineering of the Knowledge Society

Jarmo Viteli

Director of eTampere, professor of Hypermedialab, University of Tampere

jarmo.viteli@etampere.fi; www.etampere.fi

Abstract: The *e*Tampere initiative is a five-year development project that seeks to promote the development of the Information Society through measures targeting the following focus areas: the availability of public online services to residents; the strengthening of the knowledge base of research and training; and the generation of businesses related to the Information Society. This paper is based on evaluations done by Nordregio and Euricur. In this paper the model of *e*Tampere will be described, how it functions and what has been achieved so far. Also the new roles of Business, Government and University (BGU) will be described that will enhance social engineering and social capital. The challenges of these types of innovative and large scale knowledge society local programmes will be also discussed.

Key words: business, government, industry, knowledge base, public services, research, social capital, social engineering

BACKGROUND

*e*Tampere was organised as a programme focusing on three main themes, ranging from technical capabilities and access to more content-related issues:

1. The availability of public online services;
2. Strengthening of the knowledge base of research and training in the ICT-related fields;
3. The generation of new business related to the Information Society.

In order to pursue these policy goals a set of modules or sub-programmes was launched that attempted to bridge the technological and social sides of the dynamism providing a basis for developing on-line services based on existing research and innovation activities. The sub-programmes included:
- Information Society Institute (ISI);
- eBusiness Research Centre (*e*BRC);
- Research and Evaluation Laboratory (RELab);
- *e*Accelerator;
- Technology engine programs;
- Infocity;
- *e*Tampere office.

We will return to each of these in the next chapter on the governance of *e*Tampere.

THE GOVERNANCE OF *E*TAMPERE

The governance model of *e*Tampere in itself provides an interesting case study of differentiated policy measures under the umbrella of a local initiative most actively promoted by the City of Tampere. The City of Tampere clearly has a special position, though its is also worth noting that the centrality and complementary role of the various sub-programmes become obvious in the analysis of financial resources available for *e*Tampere, as well as in the nature of the project activities themselves. Ideally the different sub-programmes would work in unison and address similar policy goals through differentiated means and this in fact largely seems to be the case, based on the evaluation presented here. While for instance the business accelerator sub-programme is instrumental in attracting private capital to new enterprises, the *e*Business Research Centre provides those essential settings that are required to promote IS-initiatives, which stem from both the social and the natural sciences and where such shared projects then can meet and prosper. ISI on the other hand seeks to promote research into the Information Society almost through a programmatic and broad-based research approach on the various attributes and side effects of the Information Society. What is at the core of the programme is the need to accumulate and utilise systematic and reliable scientific information on the development and the different aspects of the information society, and this is where the different sub-programmes meet. Some sub-programmes may be bodies needing such information to identify the main needs of citizens in the face of the drastic changes currently unfolding in society and in the so-called New Economy (e.g. Infocity), while others may be seeking to influence the economic trends in the IS sphere through business promotion needs (*e*Accelerator) or testing concrete technology applications (ReLab) and thus seeking to follow the latest development in technology without which the more normative and political aspects of the Information Society would not find concrete form. The way in which these different aspects come together is through the project orientation of the programme.

The organisation of *e*Tampere is influenced by the fact that the City of Tampere has such a central role, as well as by the fact that the co-operation and co-ordination of the two universities involved determine much of the organisational structure. Project culture is also strongly present and. as is the case with similar initiatives elsewhere, also in Tampere the typical programme model today determines to a large extent the programme management practice. The project cycle methodology consists of the six-step structure described below.

INFORMATION SOCIETY INSTITUTE (ISI)

The Information Society Institute (ISI) is a joint effort by the University of Tampere and Tampere University of Technology. The University of Tampere however has operational responsibility, as the unit is placed "under its wing", in line with the contract establishing the institute between the University of Tampere and the City of Tampere in spring 2001. There is a board of directors whose task it is to steer the institute, with approximately 10 members; its membership predominantly represents the two universities involved, though there are some key external stakeholders (from the business world). The central function of the Institute is to conduct and promote research on themes related to Information Society and to pursue collaborative efforts between the two universities. It also has a role in prioritising the different research themes around Information System development, especially through its flagship project selection mechanism, which allows for concentrated efforts in some selected priority themes rather than scattering the resources too thinly.

The mission statement of the institute clarifies the objective of the institute: "to promote the construction of an Information Society based on active citizenship through multidisciplinary research and development activity and through education and training" (ISI 2001).

One of the central aspects of the Information Society Institute is the Information Society Observatory (ISO), which is a research unit launched by the Information Society Institute in June 2002 with the financial support of the City of Tampere. The main objective of the Observatory is to promote and conduct systematic social research related to the Information Society and information technology. The question of scales or levels of analysis is particularly central to the Observatory's work, just as it is to our evaluation exercise. This is quite easy to understand when one considers that Information Society, just as with ICT and technology development, not to mention policy instruments addressing citizen participation, are necessarily local, regional, national and European, as well as global (though more indirectly in evaluation terms naturally).

The ISO has a layered structure divided into three main components: education, research and network, where the network is the interdisciplinary national network of researchers representing several institutions from various geographical locations and with different research interests. The ISO promotes the co-operation of these units by developing forms of co-operation such as knowledge exchange, by creating joint funding applications and organizing meetings in scientific seminars. In addition to the networking aspects of ISO's work, it also engages in education and knowledge dissemination activities, organising masters-level thesis seminars in co-operation with a number of different University departments from the faculties of social, information and administrative sciences, as well as seeking to identify and develop both the more traditional means of knowledge dissemination, such as printed publications, books and journal and novel forms of communication such as web-portals and e-broadcasting.

Basic funding (2002): 189 000 €
Finalised projects (2002): 7 projects, with a total budget of 335 892 €
On-going projects: 35, with a total budget of 4 531 270 €

EBUSINESS RESEARCH CENTRE (*E*BRC)

The task of the *e*BRC is to bring together the two universities in Tampere, as it was co-founded by them and as such, *e*BRC functions as a joint venture that is expected to bring 'value added' to both of the partners (and to the external stakeholders).

*e*BRC seeks to turn e-business related research and development ideas into new knowledge, with its mission being outlined as:

... to generate relevant new knowledge on selected business phenomena related to the e-business, which can be utilized in the education and research activity by the partner universities and in the business practise by the businesses participating in the underlying research projects. The objective of eBRC is to become one of Europe's leading e-business researcher centres by 2006.

The method by which this task is achieved is through matching the interests of the different stakeholders from the academic community with various industrial partners, with a view to moving from domestic co-operation partners to international ones. The model has been called that of a "matchmaker" or facilitator and is also more generally typical of the kind of initiative that *e*Tampere encourages.

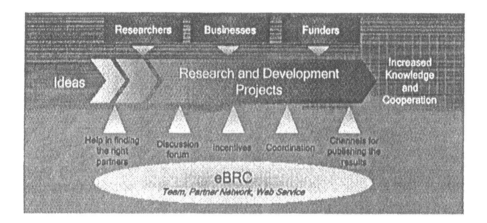

The organisational structure is also similar to the other sub-programmes, with a broad of directors responsible for strategic steering, two programme managers (one from each university) being responsible for the every-day functioning of the centre and being assisted by a small staff of project managers/assistants. The light organisational structure again here seems to be an advantage, though it does also put more pressure on key personnel, making it necessary to consider ways of improving organisational learning that would allow for knowledge to be more successfully embedded in the organisational structure, rather than relying on the personal contacts and the accumulated knowledge of the key personnel.

Basic funding (2002): 420 470 €
Terminated projects (2002): 6 projects, with a total budget of 716 973 €
On-going projects: 25, with a total budget of 2 070 541 €

RESEARCH AND EVALUATION LABORATORY (RELAB)

RELab (Research and Evaluation Laboratory for Intelligent Services) is a laboratory and testing environment situated in Tampere, which has been included as a sub-programme in the *e*Tampere programme. The responsibility for RELab lies with the Tampere premises of VTT, the Technical Research Centre of Finland. The fields of activity include services for mobile users; risk management of future services; human-computer interaction and road-mapping new and emerging technologies. The future direction of RELab lies in particular in international collaborative projects such as those within the EU's Sixth Framework Programme.

Basic funding (2002): 252 282 €
Finalised projects (2002): 2 projects, with a total budget of 269 000 €
On-going projects: 13, with a total budget of 2 274 139 €

EACCELERATOR

The *e*Accelerator Programme is located at the Tampere technology Park (Hermia), where it aims to incubate and 'hatch', by 2005, a total of some 20 companies as globally successful technology companies. The quantitative goal set consists of personnel growth in these companies to 1500 employees and turnover figure of 250 Meuro. It is the objective of the programme that each year 5-6 new companies join *e*Accelerator through the national search programme.

Basic funding (2002): 336 376 €
Finalised projects (2002): 1 project, with a total budget of 4250 €
On-going projects: 3, with a total budget of 5 859 376 €

TECHNOLOGY ENGINE PROGRAMMES

Technology engine programmes are co-ordinated by the Digital Media Institute at the Tampere University of technology. The mission of these programmes is to create knowledge and expertise in the Tampere region in the areas of adaptive software components, user interfaces, the perception of information, "neoreality" and broadband data transmission. Each field is managed by a programme manager, who seeks to create new projects in his or her area of responsibility, in close co-operation with specialists in the field working in business or research.

The goal of Technology engine programmes during the programming period of *e*Tampere, i.e. by 2005 is to have a project portfolio totalling a turnover of 19 million euros, and with approximately 100 researchers working in these projects in the Tampere region.

Basic funding (2002): 252 282 €
Terminated projects (2002): -
On-going projects: 28, with a total budget of 3 244 109 €

INFOCITY

Infocity is the one of the *e*Tampere sub-programmes that is most visible to the ordinary citizen. The mission of the sub-programme is:

...to develop the city of Tampere into a model of Information Society by offering its citizens practical online services and by improving their network skills.

This mission is implemented through the active development of digital services provided by the City of Tampere to its citizens. In addition to generating digital services with practical applications, the Programme aims at providing every resident of the City with access to these services. The Programme is an open cooperation project within the public administration as well as between the administration, private operators and the third sector. Perhaps the best-known project within Infocity's portfolio is that of the *e*Tampere City Card, which seeks to develop a multi-application, dual interface smart card system providing key Informational Society services. The project is co-ordinated by Access International Consulting Oy Ltd. VTT (Technical Research Centre of Finland) is providing additional technical consultant support. The pilot project was run during the first half of 2003 with a "test-bed " of local students. Gradually the smart card will be offered to all citizens, with the distribution goal being 100 000 units, or 50% of the city's population.

Basic funding (2002): 1 025 946 €
Terminated projects (2002): 13 projects, with a total budget of 209 767 €
On-going projects: 22, with a total budget of 1 016 696 €

As the activities of the sub-programmes vary in the pursuit of the common strategy, it is important to identify and promote priority areas that can potentially bring the sub-programmes closer together. The focus areas for 2003 have been identified as:
– Active, participating and influential citizenship (i.e. Increasing opportunities for influence and participation, eTampere card, New possibilities for communal spirit and operation).
– Developing expertise and business development (i.e. eTampere business service, Mobile technologies, services and content).
– Development of service innovations (i.e. Support for and development of service processes with information and communication technologies; Customer-oriented approach, active customer base, multifarious production).
– Increased in-depth international cooperation (i.e. Sixth Framework Programme; St Petersburg cooperation).

RESULTS OF *E*TAMPERE (NORDREGIO 2003)

The informants were asked to rate the usefulness of the *e*Tampere initiative on a scale of 1 to 5, where 1 rated *e*Tampere as being very useful, and 5 rated *e*Tampere as being least useful. On average, the usefulness rating recorded was 1,75 – a useful measure. An exception here were the *e*Tampere administrators who univocally rated the usefulness at 1. Representatives of the local and regional authorities displayed a more sceptical stance, rating the initiative as "average" in terms of its usefulness, with an average rating of 2,33.

Ranking the usefulness of various components of the *e*Tampere initiative, the respondents were harmonious in identifying "improving co-operation between local institutions and organisations" as the most enabling factor in determining the usefulness of the initiative. At the next level, there was some disagreement between the different stakeholder groups. HEI representatives identified improved national and international visibility as the second most important factor, while representatives of the local business and public sector authorities instead forwarded improved economic competitiveness as the more instrumental component:

Table 1. Most important factors for the usefulness of *e*Tampere by number of respondents (multiple choice)

Statement	Al	HEIs	Bus/Adm
Added quality and access to relevant information …	9	2	3
Improved national and international visibility	13	7	3
(Improved) economic competitiveness	13	6	6
Making Tampere into a 'model city of the knowledge society'	5	2	2
Improving co-operation between local institutions/organisations	19	8	7

In Table 1. the difference between columns 2 and 3+4 is made up of the unaccounted for responses of the four *e*Tampere administrative respondents in the sample. As can be gleaned from the table, they diverge from the priorities of the HEI and Bus/Adm respondents by putting a higher value on improved access to relevant information in electronic form.

The local stakeholders have a range of different means to meet the priorities mentioned above. When asked to rate three of the most important means, services and enabling official electronic transactions came out on top and were mentioned by half of the respondents. From a list of 14 different means no other factor was identified by more than eight respondents.

Overall, the respondents in the sample identified public interests/stakeholders as the beneficiaries of the *e*Tampere initiative compared with private interests/stakeholders on a ratio 2:1. Within the different subsets, there were natural systematic differences. HEI representatives ranked public/private interests as the main beneficiaries on a 5:1 ratio, while the business and public sector communities held a more balanced 1:1 view on who eventually will benefit from the *e*Tampere activities.

Below the situation at the end of October 2003 is summed up:
– The programme advances according to plan (project portfolio 52 million euros, realization rate 114 %);
– The proportion of finance from enterprises and international sources has increased in total funding;
– Capital investment exceeds 9.5 million euros;
– Ca. 400 researchers and more than 150 enterprises participate in over 100 *e*Tampere projects. 11 international projects;
– Information Society Observatory, which aims to produce high-quality social-scientific research on information society;
– WLAN Hämeenkatu and digital TV user studies;
– Smart home – from pilot to mass product;

- W3C – Finnish regional office in Tampere;
- 9 accelerator companies, several companies in pre-accelerator;
- SME service model –from fax-mode to *e*-era;
- 2500 people trained onboard the Netti-Nysse Internet bus, almost 20 000 visitors;
- More than 4 000 citizens trained with training funding;
- *e*Tampere card, a new key to services;
- New public online services developed. Total eservice map created for the city. The city website visited more than two million times a month;
- Several extensive international and national seminars (*e*Global, eBRF, Spearhead network seminar etc.);
- New culture of cooperation among stakeholders.

CONCLUSION

It can be argued on the basis of the evaluation presented here that, as a public policy initiative, *e*Tampere has been successful in implementing the selected policy measures designed to achieve its goals. Goals that were set in accordance with the wider European approach (as outlined in the *e*Europe strategy) towards Information Society development. The key goals of (1) making municipal services available online and (2) ensuring that as many of the citizens of Tampere as possible have access to them, whilst also (3) developing the aspects of Information Society other than those attached to e-government i.e. strengthening the Regional Innovation System (RIS) within the ICT sector (ranging from the R&D sector to the businesses and the public authorities involved) is well under way, though the final outcome cannot be seen as yet. There are indications that *e*Tampere has however now been successfully integrated into the wider sphere of innovation policy (e.g. based on survey results), though the clustering aspects of *e*Tampere (i.e. the creation of new businesses, business ideas and applications for Information Society development) can only be seen in a wider perspective. As long as the monitoring system is not further developed in line with other policy instruments for the pursuit of similar policy objectives, it is difficult to assess whether certain developments within the urban region or the RIS are due to *e*Tampere or to some other policy instrument addressing similar objectives.

BIOGRAPHY

Jarmo Viteli has been professor since 1999 at the University of Tampere and director of Hypermedialab (www.uta.fi/hyper). Hypermedialab is one of a leading research and education units in Europe in the area of Digital Media and its implementation in various areas including education. It has around 60 researchers, many international research projects and other cooperations world wide. Currently he is on leave from his position at hypermedialab and works as director of *e*Tampere-information society programme (www.etampere.fi). The aim is to invest 132 miljon € in five years time to develop Tampere-region as one of leading regions in the Information Society development. Work includes international cooperation and strategic planning.

Prof. Viteli is actively involved in many academic activities in the area of information society and new forms knowledge construction and education.

REFERENCES

ISI (2001) *The Information Society Institute as part of the eTampere programme: Proposal for the Institute's main action guidelines*, January 31, 2001.

Information: a strategic resource for technological change

Jean-Marie Leclerc
*Directeur général, Centre des Technologies de l'Information (CTI), Genève, Switzerland;
Tel. : +41 22 327 79 67; Fax : + 41 22 327 48 77*

jean-marie.leclerc@etat.ge.ch

Abstract: Our society, formerly industrial, is nowadays changing into an 'e-society',
              built on technology, information, communication and knowledge. This
              transition is facilitated by the tremendous developments in the Technology of
              Information and Communication (TIC). In this 'e-society' a new model of
              organisation is developing: the classical hierarchy is changed into a network
              designed organisation. From an economic viewpoint companies and the state
              are aiming at the maximisation of their profits, privileging cost-effective
              activities and reducing their cost. From a social viewpoint the present social
              system is more based on economical subsistence than focused on the essential
              needs of individuals. This results in serious social troubles, such as
              protestation, inequalities, and potentially aggressive frustrations. Exclusion is
              increasing in micro-societies as well as in cultures. The spread of new
              technologies produces a gap between those who can contribute and those who
              are left behind, creating the risk of cyber-exclusion. Only a sustainable
              development approach in governance will be able to balance the real needs of
              the citizens with respect to social, economic and environmental issues.

Key words: communication, cyber-exclusion, e-Society, information, knowledge, network
              organisation, sustainable development, technology

GOVERNANCE FOR HUMAN SUSTAINABLE DEVELOPMENT

In order to reach a more human 'e-society', in spite of its mechanical nature, one must notice the importance of societal indicators on the charts of governance institutions. According to the United Nation Development Program (PNUD) not only states are involved, but private sector and civil society as well. Social change funds, as well as technological expertise, are essential elements. However, postulating a planetary technological revolution implies implementation of Information and Communication Technology (ICT) at each level of the population and in every location, compatible with every political system. Nevertheless, it is only through individuals that technology can be socially implemented and integrated, thus gaining credibility.

Society consists of institutions, relations and norms which determine the quality and the quantity of social interactions. Social capital dynamically offers a common space for groups and communities, contributing to inter-resolution of problems. Social capital is not only the total of societal institutions, but also the cement which knits them together, influencing the intensity and the quality of social cohesion.

Social cohesion not only results in social values, but also significantly contributes the economy. It is an essential factor in enhancing sustainability as well as long term development. Moreover social cohesion is a guarantee for good circulation of information, communication and knowledge. These are the keys for success and autonomy in the new networked situation. In order to successfully switch to an 'e-society' based on sustainable human development governance, it is necessary to create network organisations using strategies focused on individuals.

To social and human capital organisations must, in the first place, apply principles of management based on solidarity, complementarity and balance. In the second place they should heighten civil society awareness about the role of ICT.

INFORMATION TECHNOLOGIES AND OPEN COMMUNICATION

'E-society' is not only focused on individuals, it also depends on the coordination of States, private sector and civil society, especially with respect to ICT performance. Information systems must be accessible, and therefore open. Moreover they should also smoothly fit organisational processes and guarantee satisfaction of identified needs, sustainability and confidentiality. UNESCO, the United Nations Educational, Scientific and Cultural Organisation, recommends development of Open Source software in order to enhance solidarity, cooperation and collective work among developers and users of new technologies. When built on Open Source software, information systems favourably contribute to the transition toward an open 'e-society' through wide networked accessibility, thus producing social capital.

Technology indeed has two faces: a rather exciting one, as well as a threatening one. On the one hand technologies extend the human potential with respect to calculation, simulation, communication, etc. But on the other hand the outcomes of their use are quite uncontrollable. To be realistic, not driven by an illusion of omnipotence, nor hindered by existing boundaries, this ambivalence must be part of a comprehensive approach of information management. This implies that it is the current mission of States, private sector and civil society to integrate differences in pace, funds and priorities. On the one hand 'Cyber-inclusion' is the only way forward, while on the other hand critical observation must lead to efficient risk management.

'CYBER-INCLUSION'

The definition of exclusion is logically connected to that of inclusion. Indeed, criteria of inclusion define de facto reasons for exclusion. But this is not sufficient to explain the issue. Powerful majorities, by lobby or by consensus, also play a role. Also, next to economical and social criteria, moral and ethical criteria also enlarge the number of constraints in the system. Through its integrating power every society logically also produces exclusion. Unfortunately these exclusions are reproduced by education, history, believes and moral code. Modern society does not escape this perverse cycle: barbarous exclusions are to be encountered on account of race, health abilities, sexual orientation, nationalism, employment, etc. If no action is taken, our modern life, based on technologies and information science, will produce firstly numeric and secondly economical exclusion.

The current challenge is not to eradicate exclusion and injustices all at once, but rather to benefit from world wide communication and networks to bring specific patterns to light. For 'e-society', as a sub-ensemble of common society, the same rule applies: injustice will go on. The main concern should be: *how to intervene so that 'e-society' will not add a new type of exclusion, but will rather allow a new type of inclusion for those previously excluded by the criteria of other sub-ensembles.* 'Cyber-inclusion' can be promoted in three domains:

1. *'Cyber-inclusion' by technology*: sufficient technological devices (computers, software, Internet networks, etc.), as well as legal structures guarantee access to 'e-society' for individuals and more general players.
2. *'Cyber-inclusion' by knowledge*: as part of a dynamic and comprehensive learning process, ICT is accessible to everybody at different times in life.
3. *'Cyber-inclusion' by activity*: a "three dimensional information approach" with three types of activities:
   a) Necessary activities: activities which are needed by social and economic spheres in order to be reliable;
   b) Requested activities: outcome of demands expressed by potential clients, ready to give a mandate;
   c) Feasible activities: human capital able to respond to requested activities as an efficient means to realise entrusted mandates.

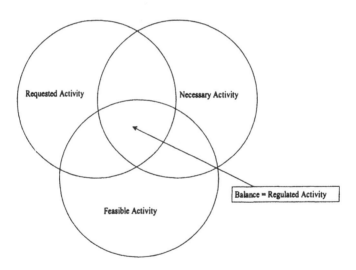

Figure 1. Activities in the three-dimensional information approach

The intersection of the three circles in Figure 1. is the zone of best balance between the various activities, containing 'regulated activities'.

Today, feasible activities are in general those that produce maximum profit. However, the volume of requested and feasible activities is constantly decreasing. An immoderate level of maximisation of profit can only be attained by a small number of people who have the level of requested competences for its realisation. This increases the number of excluded, while a few get tremendous profits.

However, 70% of current activities are not yet transformed into requested and feasible activities, because these do not allow maximization of profits. These activities offer the possibility to contribute to social security by integrating into the economical circuit people who were previously excluded. Financial autonomy, as well as added value to society, is the valuable result of this new type of contribution.

TECHNOLOGICAL CYBER-INCLUSION

The integrating power of ICT becomes very clear in the light of overcoming handicaps. Not only does the technology extend our capabilities in communication, calculation, programming and simulation, but equally execution of tasks and interconnected teamwork are in the reach of all, through simple adaptations on the logistic level. A Braille line connected to the computer is an example of how simple inclusion can be extended to the blind population.

Open Source software contributes significantly to building networks that function in human sustainable governance mode. Concretely speaking, technological 'cyber-inclusion' is providing services to civil society, such as access, knowledge and activities which here are considered as the three main elements.

CYBER-INCLUSION THROUGH KNOWLEDGE

According to the Organisation for Economic Co-operation and Development (OECD) human capital can be defined as: "knowledge, qualifications, skills and other qualities possessed by an individual". Therefore, the more individuals enlarge their human capital, the broader and the richer 'e-society' will be. Contrary to a tendency among some players in our society to keep back information and keep competence low 'e-society's dynamics' totally turns this around. The first universal skills are information and human capital management skills. Flexibility in dealing with information and abstractions is one of the fundamental competences for participating in the 'e-society's activities'. Such qualification should be developed in the economic and/or civil society sphere.

CYBER-INCLUSION THROUGH ACTIVITY

Every individual has the right of personal expression in activities for developing one's potential. On the other hand everyone should contribute to the generalisation and recognition of this particular right. The challenge of cyber-inclusion is technological development and inclusion of minorities, such as handicapped people and marginalised lifestyles, in parallel. Least developed countries must also be integrated into this planet wide change. This can mainly be done by searching the appropriate environment for the target population, based on their established potentials, and then to foster emergence of networks.

Three fundamentals principles are assisting in this process: integration as foremost concern, the three-dimensional approach of activity harmonising the implementation and thirdly, translation into action through adequate structures, promoting cyber-inclusion as well as flexibility of networks.

THE E-SOCIETY REPOSITORY

As mentioned before the current mission of states, private sector and civil society is to realise the integrating and developing process, dealing with differences in pace, funds and priorities. Additionally, critical observation has to be included to manage the various risks. Information society has a significant strategic dimension. To coordinate, diffuse and value information are actions which are necessary to regulate and to back up any technological change. An e-society repository can help in successful implementation as a tool, not only to build and follow projects, but also as road map including forgotten aspects such as ethics and coordination.

Two illusions threaten success. The first is the wish for absolute control. As we are speaking of a planet wide change, the manifestation of this illusion can take unlimited form and expression. On the other hand local adaptation and individual assertion are normal phenomena. The pay back is an extended market, based on generated activity. Rather than taking advantage of monopolisation of technology, information technology that is everywhere equally implemented, will create much broader markets and much more prosperous opportunities. This is in line with the insatiable request for multi-lateral exchange, expressed internationally as well as nationally. The second illusion concerns transparency in terms of strategy. In the common race to success and profit the most up to date strategy is an absolute necessity. However, surprisingly only a common vision can guarantee individual existence. Globally thought, locally implemented and individually expressed, that is the key to combine natural development and electronic contribution.

In this the notion of e-repository becomes a common message, subsequently translated into local action. It acts as a technological observatory, a solution reservoir, where analysis, expertise, knowledge and competence can be regrouped, shared and capitalised. In this way the strategic component of information stays open and accessible. In the end open communication and ethics reduce the threat of uncontrolled creation, which technology sometimes brings about. Thus human sense is preserved without market and environment constraints.

HUMAN BEING AS MAIN CONCERN

Social cohesion as well as social capital can be the result of this cyber-inclusion approach, but only if network organisations are developed and empowered. These seem to provide the best environment for developing human capital through activities and through exchange of knowledge and skills.

Networks, as boosters of human capital, are the answer to problems with subdivided and centralised power. The classical pyramidal (hierarchic) organisation slows down initiative and reduces autonomy. On the other hand a network organisation fosters transversal relationships and especially vertical and transversal circulation of information. Withheld information is therefore a loss of potential. The e-approach prevents subsidiary circulation of information, which often separates the monitoring process from the activity process. The raw material of 'e-society' can therefore be defined as knowledge and skills. Thus, a human network structure delimits the framework of action for "e-activities". Inside this framework, autonomy is highly recommended and social cohesion, through interdependency, is preserved by interconnection and exchange of information.

Human concerns also integrate economical constraints, as trade is a major key to sustainable development. Therefore the extension of human capital, through networks and cyber-inclusion, should not be perceived as a "charitable attitude", but rather as the only rational cold approach to remain competitive in e-business and e-trade.

CONCLUSION

The current challenge is to create a new vision of technological potential and value, taking into account changes requested by civil society. Respecting economical constraints, human development can therefore "surf" on the natural tendency of technological advancement. The evolution of ICT brings world wide communication within reach and creates consequently not only a comprehensive awareness, but also insight into proximity through connexion.

Parallel to these innovations, society is being artificially divided into those who have this particular access and those who are deprived of it. Two regulations are possible by the environment: either "natural selection" or coordination of the various players. States, private sector as well as civil society must be actively involved, so that creation of markets and new activities can be promoted and developed. This is more a strategy of autonomy arranged by states, civil society and private sector, rather than competitiveness orchestrated by profits.

The professional sphere is switching to network organisation in order to break the hierarchic shell. The use of technology in society shows that machines are easily integrated and assimilated: technology is thus not a danger for social balance, but rather a tool for development. Therefore, to remain sustainable, markets must aim at adaptation and flexibility focused on individuals and take into account human variables. Rather than retain progress and monopolise information management, technology must work for cross-cultural and world wide communication, potentially spread out to everyone.

In conclusion, the profound change from industrial times to he information era must be supported: socially – through cyber-inclusion, economically – through creation of new markets, and with equity – through multi-lateral development projects. Consequently new technologies will not be a threat of social disturbance, but rather a tool for autonomy and flexibility of individuals.

BIOGRAPHY

Jean-Marie Leclerc is director of the Centre of Information Technologies (CTI) in the State of Geneva (Etat de Genève) since March 2001. He gained his PhD at the University of Compiègne, Paris, France and his thesis was dedicated to the conception of information's systems in decentralized and multi-disciplinary environments. He has international contact, particularly in the field of development of information science related to health systems. He is lecturer at the University of Neuchâtel and several Swiss schools for engineers in telecommunication and management. He is active in the development of multilateral exchanges among public administrations, especially in the region of Neuchâtel (inter-administration project), Switzerland.

Preserving information

André Hurst
Professor of Greek, University of Geneva, Switzerland, 24 rue Général-Dufour, CH-1205 Genève, Switzerland

Andre.Hurst@rectorat.unige.ch

Abstract: The decipherment of linear B writing in 1952 by Michael Ventris has thrown light on a very interesting issue: in the second millennium B.C., the Greeks had a writing system, but the amount of information they put into writing was limited. Other civilisations, including the Celts, made the same choice. A choice was made between what you memorise and what you write down; the reasons for this choice could be political and religious in nature. In the course of history, there has been often a great resistance to writing, a resistance which in most cases arose out of a form of respect for the human memory, and a strong diffidence towards storing important information outside the human brain. This resistance is certainly connected with the status of sacred object that has often been bestowed on specific writings. It is interesting and curious to observe how the problem of preserving information outside of human memory has appeared again with the introduction of computers.

Key words: information science, memorisation, preserving information, writing

WHAT SHOULD BE PUT IN WRITING?

One of the major scientific breakthroughs of the twentieth century in the human sciences was the decipherment in 1952 by the Englishmen Michael Ventris and John Chadwick of what was called the "Minoan script" from Crete. It was perceived that there were in fact several scripts, and that the only one which transliterated texts comprehensible to us was the one classified under the name of "linear B". It transmits texts to us dating from the thirteenth century B.C., written in Greek.

Now, it very soon became inescapably evident that the material inscribed on the tablets was of a purely administrative nature. This brings us back to the very origins of writing: the earliest specimens to which we give that name are clay spheres from Mesopotamia, carrying inscriptions dealing with the management of flocks and herds. Thus, when people first thought of writing, it was because they needed it for their economic and administrative activities. In other words, the information that was to be stored by this means underwent a process of selection.

Nevertheless, within the "administrative" texts that thus became readable again after 32 centuries, we can discern in the Greek wording rhythmic formulae which are segments of verse of the "Homeric" type. Thus a list of "oarsmen proceeding to Pleuron" (Pylos, An 1) contains the item "e-re-ta, pe-re-u-ro-na-de, i-jo-te" (each sign in the graphic system used represents one syllable composed of consonant plus vowel, so it is a "syllabary"), which gives the rhythm $\upsilon \, \upsilon - - - \upsilon\upsilon - -$ (where υ represents a short syllable and $-$ a long one). Metricians recognise there the ending of a dactylic hexameter; so we have a poetic form within an administrative text. This is not an isolated example.

What is more, the descriptions of objects, as presented in the Homeric poems on the one hand and the administrative inventories from Pylos on the other, show an analogy of sequence (the presence of a similar "algorithm"). To see this we need only compare the description of objects in the Ta series from Pylos with descriptions such as that of the armour of Agamemnon or the shield of Achilles in the Iliad.

So we get the impression that those who wrote, had in mind something that they did not put into writing, but which was similar in its form to what was later to be put into writing under the name of Homeric poems, and that we catch it as it were by stealth. At any rate, there is interpenetration between the world of the scribes and that of the narrator-composers of epics (the "bards").

"OUR ANCESTORS THE GAULS" AND THE ORAL TRADITION OF HOMER

The practice of the Gallic tribes, as described for us by Caesar, was to use Greek characters for their administrative records (Gallic Wars, 1.29), while at the same time their "druids" (wise men) refused to entrust to writing what they held essential: they memorised it (Gallic Wars, 6.14). Thus the Helvetians, a Gallic tribe, had set out on a trek in 58 B.C. and wanted to cross the bridge over the Rhone at Geneva, taking with them inventories of personnel and equipment written in Greek characters (Gallic Wars, 1.29) and motivated by political-religious considerations that they kept inside their heads and which therefore to this day remain mysterious to us. The Homeric poems, too, were entrusted to memory; they date from times preceding the advent of alphabetic writing in Greece. If we say, for example, "Ulysses" *for the name of the hero*, but "Odyssey" *for the title of the poem telling the tale of that specific hero* (and not "Odysses", *which would sound logical for the hero of "Odyssey"* or "Ulyssey", *which would be the expected form for the title of a poem about Ulysses*), it is because the name of that hero, conceived in a language that was not Greek, contained a sound that the Greeks did not use (the "apico-alveolar flap") and which they transliterated variously as "d" or "l". This tells us that the personage in question did not have a Greek name, that he was a figure taken over by the Greeks, but preceding them, and that his adventures were long preserved in memory before reaching them. Further, the Homeric poems were composed by a process that operates by combining pre-existing formulae (one cannot help thinking, mutatis mutandis, of what in music we call improvisation, in particular in jazz or the playing of a figured base). Those word sequences are memorised as a sort of "poetic diction" with the help of which narratives can be constructed; the antiquity of some of those formulae is demonstrable, particularly in certain cases where the Greek wording must be given back its "Mycenean" morphology if it is to be correctly scanned.

French too has vestiges of old formulae in stories transmitted orally (e.g. the well-known "tire la bobinette et la chevillette cherra" from the original of "Little Red Riding Hood"). These processes were first studied in connection with Slavic oral literatures, and it is the American scholar Milman Parry who must be credited with having turned them to account in the context of Homeric studies, specifically in a doctoral thesis defended in Paris before the French linguist E. Meillet (L'epithète traditionnelle dans Homère, Paris 1928). The most ancient and venerable works in our European literature were thus composed orally and transmitted for centuries in people's memories. The information stored in this way has been aptly defined as the "group oral encyclopaedia".

For indeed, these compositions were not intended solely for the pleasure of the listeners to whom they were recited, but were in a way "designs for living", providing models of behaviour, ethical lessons, and even algorithms for operations such as launching a boat or organising a feast. The "bard" or composer-narrator was consequently responsible not only for the memorisation of formulae and of narratives, but also for the processing of the information thus memorised. The Homeric poems show us in action bards who are asked to sing about this or that subject (see Books I and VIII of the Odyssey), and the Balkan guzlars observed by Milman Parry did the same, choosing both the subject itself and the appropriate length of the recital according to the time available. This responsibility both for memory and for the uses of memory carries a dual implication, religious and at the same time political.

In the religious sphere, the reference made by the Homeric poet to a "Muse" (and the Muses are the daughters of "Memory") conferred on him the awesome privilege of communicating a vision of the ordering of the world that would be invested with a sacred aura. Thus Herodotus, writing in the second half of the fifth century B.C., would be able to say that Homer and Hesiod had given the Greeks the information they possessed about the gods (2.53).

In the political sphere, the choice of information was to be conditioned by the interplay of rival forces. Thus the famous "catalogue of ships", in Book II of the Iliad, detailing which Greeks were in the field against Troy, underwent manipulation by cities anxious to have their names feature in it, rather like people who consider it an honour to have had ancestors among the crusaders. The same claim to antiquity and, therefore, to legitimacy of rule is implicit in another Homeric catalogue: that of the ghosts of women from former times whom Ulysses meets in Book XI of the Odyssey: the loves of those fair mortals with gods served in particular to mark the starting points of genealogies that proved royal descent.

OPPONENTS OF WRITING

Given the way memory was exercised, it is understandable that the intrusion of writing, specifically a means for committing information to an object external to human memory, should have called forth objections. While the preservation of information might thereby be facilitated, the capacity for memorisation of the human individual would be impaired and flexibility in the exploitation of the information would suffer. For us, Plato is the chief objector to committing things to writing. There must have been many others before him, considering the polemical tone often assumed by the first Greek prose writers and their critical view of the order conserved in human memory.

Two passages from Plato are important in this context. First there is the account (Phaedrus 274c1-275b2) of how the Egyptian king Thamus refused the gift of writing: the god Theuth (sacred name of the ibis) has given human beings arithmetic, geometry, astronomy and various games, and finally offers them writing, presented by the god as "the elixir of knowledge and memory". After due thought, Thamus refuses, saying: "... this invention will produce forgetfulness in the minds of those who learn to use it, because they will not practise their memory. Their trust in writing, produced by external characters which are no part of themselves, will discourage the use of their own memory within them. You have invented an elixir not of memory but of reminding; and you offer your pupils the appearance of wisdom, not true wisdom ..." (trans. Harold Fowler). Then there is a famous passage in his seventh letter (344c) in which Plato argues that committing it to writing deprives knowledge of the necessary dynamism of reflection: "... after much effort, as names, definitions, sights, and other data of sense are brought into contact and friction one with another, in the course of scrutiny and kindly testing by men who proceed by question and answer without ill will, with a sudden flash there shines forth understanding of every problem, and an intelligence whose efforts reach the furthest limits of human powers. Therefore every man of worth, when dealing with matters of worth, will be far from ... committing them to writing." (trans. J. Harward).

A slight digression on this point. The position taken here by Plato perhaps reflects, apart from his political wrangles with the princes of Syracuse, an aspect of the Greek language that Goethe put his finger on when he contrasted Greek with Latin: he saw Latin as a language designed to operate through nouns; and nouns present the world in a settled, as it were docketed form. Greek works preferably through its very comprehensive system of participles; and participles express a moving, evolving reality. Writing can be seen from that viewpoint as deadening in that it fixes information.

Of course, the alphabet was not invented by the Greeks to put into writing the Homeric epics as vehicles of a people's culture. The fact that they borrowed the names of the letters from the Phoenicians, who used a consonantal notation system from which the Greeks derived the present notion of the alphabet, points rather to trade rivalry as the motive force. The fact remains that it is to the same period (between the eighth and sixth centuries B.C.) that we assign the first inscriptions written with the alphabet and the commitment to writing of the Iliad and the Odyssey.

The perverse effects of putting in writing matter with a political or religious purport can no longer be gauged by these examples. To do that, we must look to the Africanists. In a study published in 1952[1], Laura Bohannan describes how the British administrators of Nigeria settled differences that arose among the Tiv population by referring to genealogies that were transmitted orally and to which the Tiv appealed to establish their rights (just like the catalogue of women from former times in the Odyssey). To stabilise the situation, the administrators had these genealogies set down in writing. A generation later, to the great surprise of the English magistrates, the Tiv, still clinging to the oral tradition, refused to recognise the validity of the written version. In the event the collective memory had continued gradually rearranging the facts in response to changing circumstances; the written memory had accordingly become null and void.

While the perverse effect was observable, a beneficial effect was nonetheless kept in view. It was precisely at the end of the seventh and beginning of the sixth century B.C. that the Athenian statesman Solon put the laws in writing and displayed them to the public. Such direct access to the legislation constituted a guarantee of equality for the citizens. Solon was to be admitted to the company of the "Seven Sages", and it was another of those sages, Phidon of Argos, who fixed the value of money. The objective was the same: money guarantees equality for all in the conditions for acquiring material goods (witness the reserved-access shops which are a means of getting around that equality in favour of the least disadvantaged), and writing displayed for all to see guarantees equality of treatment, or at least the desire felt for it.

Note, all the same, that a certain prejudice persisted against committing to writing what could be learnt by heart. That is why we have no written works at all either by Socrates or by Epictetus, and why, in another part of the world, we narrowly missed having nothing by Lao Tzu[2].

AN IRREVERSIBLE STEP; CONFLICTING FEELINGS

Nevertheless, like it or not, whether or not one had reservations, a new chapter in the history of our species opened when it became possible to transfer information from our memories into objects, i.e. to store information outside our organisms.

---

[1]   Cited by Marcel Détienne, *L'invention de la mythologie*, Paris 1981, pp. 78-79.
[2]   If we are to believe the anecdote handed down by tradition, cf. Lao Tzu. *Tao Te King, le livre du Tao et de sa vertu*, new translation followed by a commentary on the teachings of Lao Tzu, Jean Herbert and Lizelle Raymond (edd.), Lyon, 1951, p. 8.

Litterae, "letters", still served in Cicero's time to designate the entire corpus of knowledge (incidentally, that is the original of the French "*Faculté des lettres*", from which other university faculties gradually became separated, just as in German the concept of "*philosophische Fakultät*" is based on a Greek view of philosophy as embracing the whole range of knowledge). Hence to inscribe letters was to preserve knowledge.

Once the machinery was in motion, attitudes towards resorting to the device would vary widely according to whether or not this means of storage was seen as partaking of the sacred or as a weapon of the secularly minded.

In his magnificent treatise on Chinese hieroglyphs, Jean-François Billeter tells us that writing does not have the same sense for the Chinese as for us[3]. In the preservation of information, hieroglyphs go further than letters: through their actual patterns they convey a comprehensive vision of the world as well as, by their juxtaposition, transmitting textual messages. This perhaps explains the contempt displayed by scholarly Chinese for the alphabet of the Koreans.

The Egyptians seem in fact to have been the first to be able to choose between an alphabet and a hieroglyphic system[4]. They deliberately chose the system that kept writing away from the masses and reserved its use for those able to grasp the vision of the world implicit in the signs themselves. At that price, storing information outside the human organism was perhaps perceived as politically (or religiously, which may come down to the same thing) less risky.

Conversely, we see the Greek alphabet as having been designed from the start to be a democratic, secular instrument, altogether in line with its commercial, non-aristocratic origins. Alongside Solon and his putting into writing of the laws for public display, we may cite the early Ionian "logographers", such as Hecataeus of Miletus, who turned out prose as a new product made possible by writing (in oral communication it is verse rhythm that serves as the preservative agent); and those logographers did not miss the chance of using this new device to criticise the lore transmitted through oral culture, such as mythical narratives and the genealogies of the characters. Here, the preservation of information by writing is the end result of a screening operation whose secular dimension is immediately evident. It is to some degree true that even the putting into writing of texts transmitted orally like the Homeric epics exposes them to critical scrutiny: they become accessible to all and their interpretation is no longer reserved for a professional caste. It is doubtless no coincidence that it was also the sixth

---

[3] Jean-François Billeter, *L'art chinois de l'écriture*, Geneva 1989.
[4] Giovanni Garbini, *The question of the alphabet*, in the volume containing contributions by various authors: *The Phoenicians*, Venice 1988, p. 89.

century B.C. which, with Theagenes of Reggio and Pherecydes of Syros, saw the first "allegorical" interpretations designed to defend the Homeric epics against the criticisms levelled at them.

Nevertheless, even the alphabet underwent elevation to sacred status, evidenced firstly in the fact that the order of the letters remained immutable. Where the Sanskrit grammarians had long since arranged the syllabograms representing their language in a sequence dictated by pronunciation, beginning at the base of the throat and finishing at the lips (from the back to the front of the human phonatory apparatus), the users of the Phoenician system out of which the Greeks made the "alphabet" (from the names of the first two letters) never dared tamper with the sequence of those signs. That sequence bears no relation to their pronunciation.

It was in 1978 that a specialist in oriental astronomy, Alessandro Bausani, proved that our alphabet is in origin a calendar[5] in which the letters "aleph" (1), "tet" (9), "ayin" (16) and "tau" (22) indicate respectively the autumn equinox, the winter solstice, the spring equinox and the summer solstice. Such an explanation presupposes, according to Bausani, a situation where the autumn full moon is close to the Pleiades (as was the case in about 1600 B.C., when the first sequences of signs that were to become the "alphabet" appeared) and a region where summer was a hostile season. The absence of signs between the summer solstice and the autumn equinox indicate absence of activity at that time of year: the letters symbolise the passage of time in the sequence of labours and religious festivals. Thus it seems that our alphabet too is sacred in origin and constitutes a sort of algorithm.

With that knowledge, it is less surprising to find in *Revelation* (1.8, 21.6, 22.13) the famous sentence, "I am the Alpha and the Omega" (i.e. the first and last letters, for the characters added to the Phoenician system displaced "tau" from its position as final letter): the divine figure expressing Himself in those terms is indicating His presence at the beginning and end of time simultaneously, and that pronouncement is based on a conception of alphabetical writing as sacred in itself.

This conception of alphabetical writing as sacred was already implicit in the fact that the letters of the alphabet are presented in acrostic form in the psalms of the "First Testament" (formerly called "Old Testament": 111, 112, 119, 145) and in the *Lamentations of Jeremiah*. After the famous passage in *Revelation*, Christian poems were to use the alphabet in acrostics (there are some in the *Codex des Visions* of the Bodmeriana Library at Cologny, Geneva, published in 1999). Origen even used the Greek name of the letters ("elements") to designate *the* sacred text, the Bible (Migne PG 13, col. 608).

---

[5] *Ibid*, p. 102.

Be it noted that we too in referring to it, whether or not we are believers, continue to say simply "the scriptures".

It was therefore with a range of nuances from the most sacred to the most profane that writing was perceived as a means of preserving information outside our memories.

THE NEXT STAGE IN INFORMATION SCIENCE

In conclusion, let us turn briefly to modern information science. In computer terminology, "memory" means capacity to store information in a new form of writing (even though, for most of us, that writing remains directly dependent on the invention of the alphabet). The objections we have seen raised, here and there, to the introduction of data-processing techniques are amusingly reminiscent of the resistance encountered by the introduction of writing, and that is of course why I have dwelt somewhat upon that episode in the cultural history of humanity. One might go so far as to say, without excessive irony, that feelings for the sacred and the profane are not always lacking in those who practice data processing in any capacity whatsoever: I am reminded of the short story by Isaac Asimov entitled *The last question*, his own favourite work, in which God Himself turns out, in the last analysis, to be a computer.

But there is certainly a major difference: memory in this context is quite distinct from its exploitation. By means of exploitation programs, of software whose capacity sometimes far exceeds our own, we have learnt to duplicate and extend what the synapses do in our brains. Thus, while the challenge is today, as it was in the transition to writing, to project outside ourselves powers that are within us, the objective will henceforth be far more visibly to achieve thereby a multiplication of the possibilities thus offered. The risk is probably worth taking, even if the exercise of our memories must thereby suffer.

For that matter the opposition, on which Plato takes his stand, between the memory within us and a memory entrusted to an external object might come to be considered outdated in the light of studies bringing together information science and biology, on the lines of those being conducted at the EPFL by Professor Daniel Mangue, but for the moment that is in the realm of speculation.

(Translation by J. Fraser)

BIOGRAPHY

André Hurst is professor of Greek and currently rector of the University of Geneva, Switzerland. His field of research and teaching includes Mycenaean Greek, ancient epics (Homeric poems, Hellenistic poetry, early Christian Greek poetry), ancient theater and papyrology.

André Hurst has studied in Geneva, Rome and München. He was visiting professor at McGill University (Montréal, Canada), Université de Lausanne (Switzerland), Universitatea Babes-Bolyai (Cluj-Napoca, Romania), Ecole Normale Supérieure, rue d'Ulm (Paris, France), and a member of the Senior common room of St John's College, Oxford. He has been chairman of the Board of Trustees of Conservatory of Geneva several times. He was dean of the Faculté des lettres at the University of Geneva from 1986 to 1992 and he presided the commission of postgraduate studies of the faculties of Arts of the French speaking Swiss universities until 2003.

Towards an Indigenous vision for the Information Society

Kenneth Deer & Ann-Kristin Håkansson
Indigenous Media Network, Box 1170 , Kahnawake Mohawk Territory,QC J0L 1BO, Canada.
Tel: +1-450-635-3050, Fax: +1-450-635-8479

kend@easterndoor.com, www.indigenousmedia.org

Abstract: The very concept of the *"Information Society"* is a cultural expression, originating in the context of the evolution of the industrial into a "post-industrial" world. Accordingly, its core elements - knowledge, information, communication and *Information and Communication Technologies (ICTs)* - are in fact culturally defined practices. However, the global approach of the evolving Information Society in an advancing Network Age makes these transformations a global issue. Societies with a different cultural, social and/or economic background, such as many Indigenous Peoples around the globe, are already affected by its dynamics – so far largely without being part of developing its philosophies or applications. To become truly global, and to avoid a new level of assimilation, colonization and marginalization, Indigenous Peoples must be equal partners in building the Information Society. Thus, the *"Information Society for All"* will have to embrace Indigenous concepts and visions in both its general conceptions and its implementations.

Key words: culture, Digital Divide, ethics, indigenous peoples, traditional knowledge

THE INFORMATION SOCIETY AS A CULTURAL CONCEPT

Information is processed according to an already existing body of knowledge, which defines its meaning and value for a given recipient. In other words: there is no information as such. What could be labelled as "information" is in fact as diverse as individual, social and cultural diversity. Likewise, knowledge has to be seen in these contexts. Consequently, the concept of knowledge is as diverse as the idea of information. Its content, definition and rules of application are culture-bound and relate to specific cultural values and protocols.

Communication, as a means for disseminating knowledge and information, is also shaped by and depends on culturally defined regulations. Awareness about these procedures is the pre-condition for allowing mutual understanding. Social rules or conventions for its distribution and the correct interaction of the involved actors may play an important role for the communication process. Such protocols not only determine what is viewed as the proper information flow, but also the appropriate medium of communication.

Knowledge, information and communication in Indigenous societies

For Indigenous Peoples, the generation and preservation of knowledge is intrinsically linked to a complex relationship with their respective ancestral territory and its environment. Traditional knowledge not only constitutes a system of knowledge and practices, but simultaneously provides a philosophy defining the place of humans in the entire "web of life". As such, it includes an inherent ethics or moral code for interaction between human, natural and spiritual worlds, which guides the utilization of resources for human use and application of knowledge for human purposes. Maintaining these relationships with accompanying obligations and responsibilities are at the heart of the identity of an Indigenous People.

Accordingly, Indigenous customary laws provide for the classification of different types of knowledge, proper procedures for its acquisition and sharing and for rights and responsibilities which attach to its possession (Dutfield 1999/1). Genetic information originating from ancestral territories is an integral part of traditional knowledge with related ethics and cultural protocols for its utilization.

In short, Indigenous knowledge can be described as deeply holistic, collective in nature and containing an entire worldview, including social and political relations and regulations for its use and dissemination. Some of its aspects are considered sacred and secret altogether. It is rooted to a particular place and cultural context. Stewardship, guardianship and inter-generational responsibility are important principles for managing traditional knowledge.

In this context, information can be defined as "traditional knowledge that is communicated in a certain form to a certain audience ... with and for a certain purpose" (Alonso 2003). Indigenous protocols of sharing and acquiring knowledge address questions such as appropriate actors, language, context, situation and procedures to follow. Much knowledge is gender and age specific and/or is guarded by a certain clan, family or specialist. Exchange of information and knowledge, therefore, also follows these patterns of guardianship. Indigenous approaches to information communication are highly contextual and bound to the ethical and cultural obligations related to the shared knowledge. Communication of knowledge from generation to generation takes place according to culturally defined socialisation and education processes. Often, knowledge is revealed through stories, legends, performance or ceremonies. Teachings are specific in time and place, and adapted to the respective recipient(s).

Oral transmission is generally considered the appropriate medium for sharing and communicating Indigenous knowledge and information. Fixing it through writing, taping or other means is often seen as unduly defining a part or particular view as the whole, while taking it out of context. Passing on knowledge includes an obligation of the teacher to consider whether the learner is ready to use the knowledge responsibly. Recordings would relinquish the possibility of adjusting the teaching to the maturity of the learner and thereby influencing the ethical use of knowledge (Brant Castellano 2000).

INDIGENOUS KNOWLEDGE AND THE INFORMATION SOCIETY

The Information Society is a knowledge-based economy. Companies are under high pressure for constant innovation to stay competitive in a globalised world. Access to and acquisition of knowledge and information have become a strategic competitive advantage for their economic growth and survival. Knowledge and information have turned into decisive resources to fuel the "new economy" of post-industrial societies. Information, in this context, also includes genetic and biological information.

Traditional knowledge and inventions of Indigenous Peoples are an important part of these developments, mainly utilized for product development in agri-business and the pharmaceutical industry. Indigenous cultivars like rice, maize and potatoes can be used to improve commercial food and fibre crops, for example by increasing resistance to extreme climatic conditions or disease (Daes 1993). The pharmaceutical industry on its part applies Indigenous medicinal and botanical knowledge for the development of new drugs. Knowledge shared by Indigenous Peoples on specific plants, their physiological effects and methods for drug processing can provide valuable insight for the identification and isolation of active molecules. It is estimated that using ethnobotanical information when screening plants has increased the rate of discovery of biological activity by 400-800% (Swanson 1995).

Searching and collecting biological material and traditional knowledge from Indigenous territories for commercial use, so-called 'bioprospecting', has intensified in recent years (Dutfield 1999/2). Scientific estimates indicate that Indigenous Peoples possess as much as 99% of the existing knowledge about usable species (BMZ 1997) – and thus about utilization of biodiversity. However, bioprospecting and subsequent economic exploitation of Indigenous knowledge, cultivars or other biological and genetic material originating from their territories, often takes place without authorization and consent of Indigenous Peoples or adequate benefit sharing. Indigenous knowledge becomes increasingly reduced to a mere raw material for the knowledge-based economy of the Information Society.

Indigenous knowledge and the public domain

It is evident that the public domain concept, developed in the framework of European economic philosophy, does not match with Indigenous ethics and customary laws when it comes to disseminating and utilizing knowledge and information. Western economic thinking classifies all information and knowledge that it considers shared, disclosed or generally known, as part of the public domain. As such it is perceived a freely available resource for commercial use. Non-Indigenous actors have widely applied this idea to exploitation of Indigenous knowledge and cultural expressions, for example in the context of bioprospecting. As a result, Indigenous Peoples not only lose ownership and control but in fact are barred from fulfilling existing cultural obligations in breach of customary laws. Additionally, third parties often secure Intellectual Property Rights for commercial applications derived from Indigenous knowledge and information.

From an Indigenous point of view, there might be striking similarities between the historic use of the *terra nullius* concept for seizing Indigenous territories and the current conversion of traditional knowledge into a *res nullius* by defining it as part of the public domain. A similar 'philosophy of appropriation' is connected to the notion of 'wilderness' in Western economic thinking: useful plants or an entire ecosystem and its biodiversity declared as 'wild' or 'natural' belong to the public domain, for instance as common heritage of mankind. Such 'wilderness', however, might have been managed for millennia by Indigenous Peoples, who nurtured its biodiversity through developing and encouraging species diversity. From an Indigenous perspective, the distinction between 'domesticated cultivars' and 'wild relatives' might even be meaningless. Nevertheless, the label 'wild' categorizes Indigenous cultivars, medicinal plants and their genetic information as *res nullius*. For Indigenous Peoples, such denial of their rights is unacceptable.

Indigenous Peoples and Intellectual Property Rights (IPRs)

Access to a vibrant public domain is the backbone of the envisioned global Information Society and its knowledge-based economy. However, at the same time, Indigenous Peoples need to assert their rights and obligations towards their traditional knowledge. Are IPRs a solution to this problem?

The IPR-concept has been elaborated within the same philosophical framework as the public domain. In fact, both are two sides of one coin. Producing a new creative work with commercial value out of public domain resources, gives rise to private property rights for the inventor or creator. The owner can seek intellectual property protection under IPR law. IPRs are an instrument to reap economic benefits from a commercial creative work by allowing its temporary removal from the public domain.

Applying this legal concept to Indigenous knowledge and creativity raises a number of difficulties. For Indigenous Peoples, "intellectual property" does not necessarily imply ownership in the sense of private property, used for the purpose of extracting economic benefits. Instead, "possessing" knowledge is linked to community and individual responsibilities, involving a reciprocal relationship with humans, animals, plants or places it is connected to. Due to the collective status of traditional knowledge, it might violate Indigenous ethics to determine a single owner (an individual or a group) as creator or inventor. Also, for many Indigenous Peoples creation is a gift. Authorship or the source of innovation may be assigned to ancestors or spiritual beings. Human ownership is rather understood as custodianship, with future generations as strong rights-holders.

Current IPR-regulations cannot assist Indigenous Peoples in preventing unauthorized release of their knowledge into the public domain with subsequent exploitation either. IPRs are temporary and solely protect the expression of ideas or their physical embodiment, but not the knowledge itself. Instead, protection of a work as intellectual property often means disclosure of relevant knowledge. Copyright protection for Indigenous songs or stories would for instance require recording. Likewise, filing a patent involves public disclosure of the invention. Consequently, related knowledge and information can be freely utilized as long as there is no infringement on a protected work.

Evidently, IPRs cannot enforce respect for Indigenous obligations towards their traditional knowledge or protection against its appropriation. Thus, adequate *Rights of Indigenous Peoples in the Information Society* must be elaborated.

THE INFORMATION SOCIETY - CHALLENGES AND POTENTIALS FOR INDIGENOUS PEOPLES

For Indigenous Peoples, two essential questions emerge regarding their participation in the Information Society:
1. Is it possible to share their knowledge and information without violating their cultural obligations and customary laws?
2. Is it possible to utilize ICTs within their cultural contexts without risking to lose their cultural identity?

ICTs are fundamental tools for the implementation of the Information Society. They determine, how knowledge and information are communicated. ICTs are, as any technology, a cultural product expressing the cultural identity of the society that has developed them. And as any technology, ICTs bring forth a new world - emerging out of particular cultural conditions and in turn helping to create new ones (Escobar 1994).

For Indigenous Peoples, these impacts can constitute a challenge, a potential or both – depending on their participation in this process and their ability to determine ICT utilization on their own terms. Cultural appropriateness of ICT applications and content, also reflecting Indigenous modes of communication, is essential. ICT use must support and enrich Indigenous cultures, strengthen their identities and improve their quality of life. Also, ICTs cannot replace traditional elements of Indigenous cultures, such as inter-generational knowledge transmission or interaction with ancestral territories. If these pre-conditions are met, ICTs might develop into a useful complementary tool. Otherwise, they may contribute to culture loss. (Indigenous Media Network 2003)

Table 1. Towards an Indigenous Vision of the Information Society (Indigenous WSIS-Position 2003)

Key elements	Instruments
Equal participation of Indigenous Peoples and recognition of Indigenous cultural approaches towards knowledge, information, communication and ICTs	Recognition of existing Indigenous Rights; Respect for treaty rights; New standard setting activities to develop adequate *Indigenous Rights in the Information Society*
Respect for Indigenous cultural protocols for sharing, disseminating and communicating knowledge and information; Protection of Indigenous knowledge against appropriation, misuse and unauthorized exploitation; Enforcement of Indigenous cultural obligations towards their knowledge	Creation of an international legal instrument with participation of Indigenous Peoples, taking into account: - the right of Indigenous Peoples to full ownership, control and protection of their cultural and intellectual property - their culturally diverse concepts and pro-provisions of their customary laws in defining the term "intellectual and cultural property" - their cultural obligations towards communicating, sharing, disseminating, using and applying knowledge - alternatives to the application of the public domain concept to their knowledge and genetic information originating from their territories - alternatives to the application of current IPR regimes to their knowledge and genetic information originating from their territories - the collective status of their knowledge - their culturally diverse concepts of ownership - a multi-generational view towards rights holders - their right to be first beneficiaries of their knowledge - culturally appropriate mechanisms of benefit sharing - their right to say "no" - adequate monitoring mechanisms Creation of similar national legal instruments and mechanisms

Key elements	Instruments
Equal partnership in building the Information Society; Contributions to the ethics of the Information Society; Intercultural understanding.	Establishment of a high-level mechanism that brings together Indigenous and non-Indigenous actors and provides Indigenous Peoples with the possibility to continuously contribute their input towards the evolution and implementation of the Information Society; Promotion of exchange between Indigenous and non-Indigenous practitioners to foster mutual understanding and develop solutions that respect Indigenous approaches in a global Information Society

Accordingly, participation of Indigenous Peoples in the Information Society needs to be rights-based. Recognition of Indigenous rights such as those related to their ancestral territories, traditional knowledge, cultural values, educational systems, languages, methods of teaching and learning, the integrity of their traditional health systems and healing practices or establishment of their own media are fundamental for enabling Indigenous Peoples to become equal partners in the evolution and implementation of the Information Society, and to develop their own visions.

Within such a framework, Indigenous ICT utilization could include:
- Preservation of Indigenous cultures and languages;
- Indigenous education, and particularly long-distance teaching;
- Indigenous language training;
- Indigenous health education;
- Health assistance for remote Indigenous communities;
- Environmental education and monitoring;
- Support for traditional ways of life, for instance: nomadic communities;
- E-business;
- Establishment of Indigenous media;
- Intercultural education and combating racism and discrimination against Indigenous Peoples.

However, not all Indigenous knowledge and cultural expressions can be digitalized and/or digitally preserved, due to cultural protocols and obligations. Also, Indigenous Peoples are deeply concerned about losing ownership and control over use of their traditional knowledge and cultural expressions, once they are displayed for preservation, educational or other purposes - and thus classified as part of the public domain by the non-Indigenous world.

Therefore it is indispensable for Indigenous Peoples to not only control and determine the utilization of ICTs in their communities, but also develop their own culturally appropriate ICT applications and content. Furthermore, from an Indigenous point of view, it may be essential to involve Elders and other traditional authorities, responsible for guarding Indigenous knowledge, into resolving these questions.

Table 2. Towards a policy on Indigenous Peoples and the Information Society (Indigenous WSIS-Position 2003)

Key elements	Instruments
Development of Indigenous approaches towards the evolution and implementation of the Information Society and establishment of equal partnership between Indigenous and non-Indigenous actors	- Promotion of Indigenous research projects to explore the aspirations of Indigenous Peoples towards the Information Society and the potentials and challenges it poses to their communities - Support for Indigenous monitoring activities regarding legislation and business practices affecting their traditional knowledge and cultural expressions
Bridging the digital divide in Indigenous areas	- Support for Indigenous studies to develop strategies of Indigenous Peoples how to bridge the digital divide in their regions on their own terms and ensure affordable access solutions - Promotion of "ICT - Elders-and-Youth" initiatives in Indigenous communities - Support for Indigenous Peoples to exchange experiences on ICT-use among themselves - Promotion of "Indigenous-to-Indigenous" co-operation
Development of culturally appropriate ICT-applications	Design of culturally appropriate capacity building instruments on ICT-use by Indigenous experts and with participation of the Indigenous Peoples concerned - Informed decision-making processes by Indigenous Peoples on their needs for ICT-utilization and on culturally appropriate ICT-applications and content - Development and implementation of culturally appropriate ICT- solutions with participation of Indigenous Peoples, taking into account high illiteracy rates and lack of command of non-Indigenous languages, while simultaneously strengthening Indigenous languages

Key elements	Instruments
Establishment of Indigenous media	- Promotion of partnerships for culturally appropriate digitalization of Indigenous languages, if so requested by the Indigenous Peoples concerned, with projects carried out under their direct control and with their approval for every step of the process - Support for Indigenous research on the effects of ICT-utilization on the survival of Indigenous cultures, languages and identities - Developing, in co-operation with Indigenous Peoples, policies and communication legislation for the establishment, operation and funding of Indigenous media and related technical infrastructure - Support for content development by Indigenous Peoples to counter racism against them

Culturally appropriate capacity-building on technical aspects, but also on the potentials of ICTs in their various fields of application will be fundamental to enable informed decision-making of Indigenous Peoples. Preferably, such programs should be designed and carried out by Indigenous ICT-experts.

However, Indigenous Peoples, no matter where they live, are affected by the digital divide. Lack of basic infrastructure including electricity and telephone services, availability of servers and ICT-equipment and/or financial resources for necessary acquisitions prevent many Indigenous Peoples from access to and participation in the Information Society. Nevertheless access solutions have to be culturally appropriate. Thus, the right of Indigenous Peoples to bridge the digital divide on their own terms has to be recognized. One possible avenue to take would be the promotion of "elders-and-youth" initiatives that would assist Indigenous Peoples in shaping their future without losing their cultural identity and in supporting the survival of their living cultures without risking their museumisation.

To reach these goals, it is important for Indigenous Peoples to share experiences on ICT-use and to conduct their own research on these issues.

INDIGENOUS CONTRIBUTIONS TO THE INFORMATION SOCIETY

Traditional knowledge is often viewed as primitive, without value or "wild". Treating it as "raw material" for the "new economy" continues this tradition.

However, human progress and particularly European development historically have greatly benefited from Indigenous knowledge and innovations. Quinine and rubber for instance, both originating from Indigenous cultures, played an important role in European medicinal and industrial development. Potatoes, tomatoes, corn, cacao or chocolate are all the result of Indigenous creative work. Even the shaping of democratic ideas was influenced by Indigenous philosophies.

Indigenous Peoples guard a major part of existing knowledge about biodiversity, while their philosophies provide holistic guidelines for its use. Today, Indigenous knowledge continues to fuel progress in medicine, agriculture and economic production, but also on the philosophical level. Lately, it has been instrumental in forming the idea of sustainable development.

An Indigenous vision for a future Information Society seeks inclusion of their *web of life* philosophy. Indigenous Peoples understand that knowledge, information, communication and technology are not context-free, but intimately interwoven with ethical obligations towards the entire web of life. Global networks and "the web" as emerging organisational principles of the Information Society point to this reality. However, the evolving vision of human organisation and interaction within all-embracing network structures so far lacks philosophical and ethical grounding.

The very concept of traditional knowledge offers a contribution to developing a holistic philosophical framework of the Information Society, adequately reflecting its network character. Recognizing its moral principles would enrich a global cyber-community with an ethical dimension on knowledge sharing. Respecting its holistic nature would provide the actors of the knowledge economy with an ethical guideline for responsible utilization of information and content. Thus, protection of Indigenous rights enables Indigenous equal participation in shaping a common future nurtured by cultural diversity.

BIOGRAPHY

Kenneth Deer has been coordinating a study on Indigenous Peoples and the Information Society to produce a position paper for Indigenous representatives to Prepcom 3 and the World Summit on the Information Society. He was media coordinator for the Indigenous Caucus at the World Summit on Sustainable Development 2002 in Johannesburg. He was selected as an Indigenous journalist by the UN to report on the World Conference Against Racism 2001. In 2000 he was chairman/rapporteur of the United Nations Workshop on Indigenous Media in New York.

He is founder of the weekly newspaper The Eastern Door (since January 31, 1992). As owner, publisher and editor, he oversaw the development of the paper from its beginnings until today where it is a well established, award winning newspaper. He occasionally writes editorials in the Montreal Gazette, a large urban newspaper.

He is member of the Native American Journalist Association and winner of a number of awards from this association, Canadian Community Newspaper Association and the Quebec Community Newspaper Association. He was member of the Board of Directors for the Quebec Community Newspaper Association from 1999 to 2001.

Kenneth is co-founder of the Indigenous Media Network, an international organization to link Indigenous journalists around the world. (www.indigenousmedia.org)

On the international level he is coordinator and often co-chairman of the Indigenous Caucus at the United Nations in Geneva. As such, he participates in the meetings of the UN Working Group on Indigenous Populations, the Working Group on the Draft Declaration on the Rights of Indigenous Peoples, the Permanent Forum **on Indigenous Issues** and other meetings. He has been invited speaker to numerous seminars in various countries to discuss Indigenous Rights.

Ann-Kristin Håkansson has been assisting in carrying out a study on Indigenous Peoples and the Information Society and to produce a position paper for Indigenous representatives to Prepcom 3 and the World Summit of the Information Society. She has 20 years of experience in development co-operation with Indigenous Peoples, project management and project evaluation. Furthermore, she has carried out a number of research tasks in this context. Finally, she has organised Indigenous capacity-building programs for Indigenous Peoples from developing countries, but also served as a resource person in such training programs. Her main geographic areas of work are Latin America and Africa.

Ann-Kristin has worked as a consultant for the European Commission, undertaking a study on the EU-development co-operation with Indigenous Peoples. The study served as background material for the elaboration of an European Union policy on Indigenous Peoples and development co-operation. Furthermore, she was in charge of a commission funded project on conducting regional studies on *Indigenous Peoples in the Development Situation* on a global scale, resulting in the production of a booklet on an Indigenous approach to sustainable development.

Since 1977 she has worked for the Saami Council, an Indigenous NGO for Saami living in Norway, Sweden, Finland and Russia. The Saami Council was established in 1956 and has a consultative ECOSOC status at the United Nations.

On the international level, she participates in meetings related to Indigenous issues, such as the United Nations Working Group on Indigenous Populations and the Working Group on the Draft Declaration on the Rights of Indigenous Peoples, the United Nations Human Rights Commission and the Commission on the Right to Development.

REFERENCES

Alonso, Marcos Matías, (2003): Independent Expert Paper, Indigenous Peoples in the Information Society. Draft Paper by the WSIS Focal Point of the Permanent Forum on Indigenous Issues. New York/Geneva; para.14

BMZ - Bundesministerium für wirtschaftliche Zusammenarbeit und Entwicklung (1997): Promotion of Indigenous Forest-Dwelling Peoples Within the Scope of the German Federal Government's Tropical Forest Program (BMZ aktuell 062), Bonn; p.8

Brant Castellano, Marlene (2000): Updating Aboriginal Traditions of Knowledge. In: Dei, George F. Sefa/Hall, Budd L./Rosenberg, Dorothy Goldin (Eds.): Indigenous Knowledges in Global Contexts. Toronto; p.26/27

Daes, Erica - Special Rapporteur (1993): Study on the protection of the cultural and intellectual property of indigenous peoples. United Nations Sub-Commission on Prevention of Discrimination and Protection of Minorities; E/CN.4/Sub.2/1993/28; para.103

Dutfield, Graham (1999): Rights, Resources and Responses. In: Posey, D. A. (Ed.): Cultural and Spiritual Values of Biodiversity. UNDP, Nairobi/London; p. 508(1); p.505(2)

Escobar, Arturo (1994): Welcome to Cyberia: Notes on the Anthropology of Cyberculture. In: Current Anthropology, Vol. 35, Issue 3; p.211.

Indigenous Media Network (2003): Indigenous Peoples and the Information Society. Survey Summary; www.indigenousmedia.org

Indigenous WSIS-Position (2003): Indigenous Peoples and the Information Society. (Draft) Indigenous Position Paper for the World Summit on the Information Society (WSIS); www.indigenousmedia.org

Swanson, Timothy (1995): Diversity and sustainability: evolution, information and institutions. In: Swanson, Timothy (Ed): Intellectual Property Rights and Biodiversity Conservation: an interdisciplinary analysis of the values of medicinal plants. Cambridge; p.9

Vulnerabilities of information technologies
Impact on the Information Society

Klaus Brunnstein
Professor for Application of Informactics, Faculty for Informatics, University of Hamburg, Germany, Tel: +49-40-42883-2406; Fax: +49-40-42883-2226

brunnstein@informatik.uni-hamburg.de; http://agn-www.informatik.uni-hamburg.de

Abstract: Faster than any technology before, Information and Communication technologies (ICTs) continue to change economies and societies in ways affecting many aspects of human life. Already after a short time organisations and individuals have become so dependent upon proper functioning of highly complex and hardly understandable ICT-systems that any deviation from "normal" behaviour may have adversary, if not damaging effects. Contemporary technologies are in many cases designed and implemented without adequate provisions for safe and secure functioning. These systems can easily be attacked, even by experimenting boys at the age of puberty. e.g. by injecting viruses and worms into The Internet, which then rapidly propagates these malicious gifts (in some cases even in epidemic amounts of up to 100 million emailed worms per 24 hours) into enterprises, offices, schools and everybody's PC. Internet communication protocols are weakly designed, and it is easy to "spoof" ones email address, to "sniff" and to intercept messages, such as transfer of electronic funds. To protect these already overly complex systems, the usual solution is to add more complexity: firewalls, antivirus software and encryption. Two types of reaction to ICT related vulnerabilities can be observed. Some users have a "don't care" strategy, especially young people who leave a data trail of personal behaviour when surfing websites with potentially interesting economic or sociological content. Other users wish to exclude all risks and follow a strategy of "don't use". Both reactions are undesirable in the Information Society. Education on how to work with unsafe and insecure systems may help to protect users from undesired side-effects of ICT work.

Key words: computer crime, malicious software, risk analysis, risk management, safety, security, theory of economic cycles, vulnerabilities

INTRODUCTION: FROM INDUSTRIAL TO INFORMATION SOCIETIES

In the history of mankind, several technologies have contributed to change economic and individual conditions and perspectives. Most technical inventions – such as horse-driven cars, windmills and ships – developed rather slowly, both the technology and the impact on society. However, James Watt's invention (1761) of the vapour-pressure driven "machine" changed economic conditions and in their wake the world order at significantly higher pace. In a first – stationary – form heavy machines with relatively small power (though large compared to human "manufacturing") produced industrial goods in what was to be the start of essential change in the production of goods. This change gained speed and impact when Watt's machines became sufficiently light and powerful to drive wagons on iron rails ("locomotion"). This allowed faster transportation, both of material resources to the fabrication sites and of products to consumers. The "Industrial Society" even gained more momentum with the application of other technical inventions such as electric energy, telephone and energy production from oil and uranium.

In 1960, when computers began to significantly spread into (then large) businesses and universities, the so called (technically) "developed countries" showed significant economic and societal changes since the advent of Watt's engine. These changes affected many areas: from hierarchical to democratic organisations, legal systems to support individual rights, opportunities for education and medical support for all, etc. But seen over the span of about 200 years of industrial history (1762-1960), the speed of change was rather slow.

Table 1. Technologies supporting development cycles of Industrial Societies

Cycle 1	(1760+)	Vapour driven stationary machine (James Watt: 1762)
Cycle 2	(1810+)	Vapour-driven mobile machine
Cycle 3	(1860+)	Oil-driven engines
Cycle 4	(1910+)	Electricity-driven engines, networks

Analysis of similarities and differences in development of industrial technologies may help to predict future developments of ICTs. Schumpeter, Kondratieff et al. have observed cyclic behaviours in industrial economies when new technologies stimulated economic growth. Looking at essential ("lead") technologies industrial developments can be classified into essentially 4 cycles, each with about 40-50 years of duration (~45 years). See Table 1.

Faster than industrial technologies before, Information and Communication technologies (ICTs) continue to change economies and societies in ways affecting many aspects of human lives. ICTs are lead technologies of the Information Societies (Table 2.). Cycle 1 of Information Societies depended on initially large, heavy and difficult to handle "engines" (mainframes and their software) operated by specialists in "computer centres" – similar to factories in the Industrial economies - which developed into smaller and more powerful machines (Mini, Micro and Personal Computers). This made it important to transport digital results between remote sites and networks developed both within enterprises (Intranets, Local Area Networks) and globally. "Traditional" communication – telephony using terrestrial cables, satellite and mobile cell-based communication – has enlarged human communication space even in its previously analogous, but now digitalized forms. Cables are still the major basis of digital networks, but in some areas flexible mobile communication is a major factor of information exchange in enterprises and daily life. The development of The Internet – designed and intended for scientific and military communication – became the major "lead" technology for the 2nd Cycle, similar to the development of railway transport in the 2nd Cycle of the Industrial Age.

Table 2. Technologies supporting development cycles of Information Societies

Cycle 1	(1940+)	Computer: Mainframe .. PC .. Chips; stationary, local code/control; computer companies support economic development.
Cycle 2	(1985+)	LAN ... WAN, mobile code/agents; data searching & mining, value-added services; network companies lead development.
Cycle 3	(2030+)	??? (Nano-miniaturisation: Quantum/Optical Computing) ???

In the Industrial age, railways enabled faster transport of resources, products and persons changing economic and human relationships as these progressed from local to regional, and finally to global activities. In similar ways Information and Communication technologies have started to rapidly and deeply affect many relationships that originate from Industrial societies.

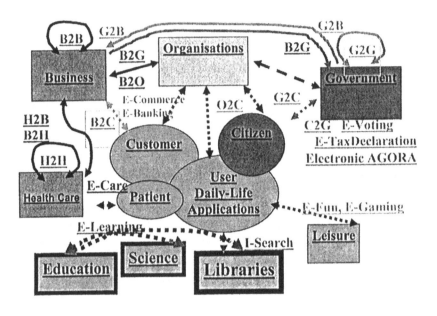

Figure 1. Relations in e-Society

Digital communication is facilitated by Local-Area Networks (LANs) and the Internet as a cooperation of many Wide-Area Networks (WANs). Based on the uniform Internet communication Protocol (IP) it is changing many relationships (see Figure 1.) in and between enterprises (Business to Employee, B2E; Business to Business, B2B), but also between business and customers (business to customer, B2C) and governments (B2G). Also, though with lower speed, individual relationships are changing: for patients (Health to Patient, H2P), for students (School to Student, S2S; Student to University, S2U) and citizen (Citizen to Government, C2G).. Essential processes of the Information and Communication Societies will be based upon production of digital values such as Information Search (Data Mining, I-Search). For example, changes in education such as distant and e-Learning, as well as in democratic processes, e.g. e-voting, are underway.

Remark: following contemporary implementation, such processes are labeled "e-" for "electronic", thus indicating that in the implementation electromagnetic media and processes are used. As future technologies will possibly also other technologies such as photonic and quantum technologies, it seems advisable to label such relations as "d-relations" thus reflecting their digital (rather than analog) nature.

It is true that speed and impact of developments in ICTs vary greatly over the planet. As in the Industrial Age, several countries (better: regions within nations) are advancing rapidly in technical developments with associated benefits, especially transfer of related products and methods to other parts of the world. The uneven distribution of regional development also implies that e-relationships develop rather differently over the planet. The "digital gap" hinders less "developed areas" to participate in this global process. On the other side, this "digital gap" protects less-developed areas from unwished side-effects and serious (e.g. security) problems of these technologies which affect enterprises, governments, organisations and individuals with undesired effects.

VULNERABILITIES OF INFORMATION AND COMMUNICATION TECHNOLOGIES

Already after such a short time organisations and individuals have become dependant upon proper functioning of these highly complex and hardly understandable systems. Any deviation from "normal" behaviour may have adversary, if not damaging, effects. Daily experience of "users" of Information and Communication Technologies is that computerized systems fail, rather often for reasons that a "normal user" can hardly analyse or understand. Failures range from unforeseen crashes, infections with malicious software (in many cases imported by Intranet or Internet communication) to loss of data and programs, to complete loss of function and connectivity.

A Risk Analysis of contemporary ICTs reveals a variety of reasons why such systems fail. Following the "life cycle" of ICT concepts and products, risks can be graded into "vulnerability classes":

- Paradigmatic risks: paradigms dominating design;
- Implementation risks: deficiencies in the quality of production and products
- Usage risks: risks in the ways systems are used
- Criminal risks: in addition, ICTs may be deliberately misused for criminal purpose.

Paradigmatic risks (Vulnerability Class 1)

Risks in this class are embedded in assumptions made in the design process and in the methods applied in production (implementation) of hardware, systems and application software. Take for example the concept that complex problems can be solved by cutting them into parts ("modules") which can separately be produced, to be subsequently combined in systems of high complexity. Not only are there the problems of adequate "cutting" of modules and of designing adequate interoperation, but contemporary systems also have become so complex that even experts can hardly understand their effects. To get still more functionality and interoperability, complex systems are combined to produce even higher levels of complexity. When systems from different origin with no common "interface" are combined, instruments are needed to "glue" such systems together; such "glueware" – "script" programming languages such as Java or Virtual BASIC – must be powerful to be able to support many different adaptations, but then also can be easily manipulated even by less qualified "script kiddies". The plenitude of malicious software (computer viruses, network worms, Trojan horses, trapdoors, backdoors, spyware, etc) is essentially based on script languages used for example. in office systems (Visual Basic for Applications etc.). In summary, concepts and tools used in the design process very deeply influence both the functions and the risks of digital technologies.

Risks from inadequate implementation (Vulnerability Class 2)

The production of digital technologies, especially system and application software (and to a lesser degree also of hardware and their "drivers"), has many weaknesses. The most evident are: inadequate qualification of programmers, and inadequate testing and production under heavy time pressure. Conceptual and programming errors have effects, mostly for users, when software behaves unpredictably, including destruction of achieved work and broken connections. The number of experienced "computer emergencies" is rapidly growing, with sometimes millions of servers and even more local computers being affected by software weaknesses and "infections" by network "worms".

Effects of software weaknesses of course materialise predominantly on systems of the most dominant system and software producers. Microsoft therefore globally leads in sold software and in software flaws, but is also in the lead with malicious software living on Microsoft's insecure software design and implementation. Table 3. shows a list of programming faults, detected within last year. Although there are also many problems with non-Microsoft systems (e.g. Linuxes), Microsoft dominates in those incidents where in some cases many millions computer systems were affected.

Table 2. Vulnerabilities reported by CERT/CC (CERT/CC October 2002-October 2003)

A1	Buffer Overflow Vulnerability in Core Windows DLL
A2	Remote Buffer Overflow in Sendmail
A3	Increased Activity Targeting Windows Shares
A4	Samba Contains Buffer Overflow in SMB/CIFS Packet Fragment Reassembly Code
A5	MS-SQL Server Worm
A6	Multiple Vulnerabilities in Implementations of the Session Initiation Protocol (SIP)
A7	Multiple Vulnerabilities in SSH Implementations
A8	Buffer Overflow in Microsoft Windows Shell
A9	Double-Free Bug in CVS Server
A10	Buffer Overflow in Windows Locator Service
B1	Integer overflow in Sun RPC XDR library routines
B2	Multiple Vulnerabilities in Lotus Notes and Domino
B3	Buffer Overflow in Sendmail
B4	Multiple Vulnerabilities in Snort Pre-processors
C1	Multiple Vulnerabilities in Snort Pre-processors
C2	Exploitation of Vulnerabilities in Microsoft RPC Interface
C2.a	W32/Blaster Worm
C2.b	W32/Welchia
C3	Cisco IOS Interface Blocked by IPv4 Packet
C4	Vulnerabilities in Microsoft Windows Libraries & Internet Explorer
C4.a	Buffer Overflow in Microsoft Windows HTML Conversion Library
C4.b	Integer Overflows in Microsoft Windows DirectX MIDI Library
C4.c	Multiple Vulnerabilities in Microsoft Internet Explorer
C5	Malicious Code Propagation and Antivirus Software Updates
D1	W32/Mimail Variants (added: plus Paylap variants)
D2	Buffer Overflow in Windows Workstation Service
D3	Multiple Vulnerabilities in Microsoft Windows and Exchange
D4	Multiple Vulnerabilities in SSL/TLS Implementations
D5	Exploitation of Internet Explorer Vulnerability
D6	W32/Swen.A Worm
D7	Buffer Overflow in Sendmail
D8	Buffer Management Vulnerability in OpenSSH
D9	RPCSS Vulnerabilities in Microsoft Windows

Risks from usage (Vulnerability Class 3)

Not surprisingly, usage of unsafely designed and insecurely implemented software presents additional risks. When installation and administration of system and application software is improperly performed at user sites, this may adversely affect performance and proper functioning. Due to the complexity, as well as inadequate documentation, users hardly understand the effects of their attempts to "properly" use such systems. Consequently users apply "trial and error" methods in learning to work with new features, rather than trying methodologically to understand which functions may have which effects, and which precautions should be taken to avoid unwished side-effects. This somewhat "explorative" way of using a system rather often leads to a risky attitude, with potentially hazardous effects. For example clicking on unknown attachments without due care.

Software manufacturers often argue that failure of software is mainly caused by improper actions of users. But in many – if not most – cases, the human-computer interface (e.g. the display of functions and operations on the screen, or the handling of input devices such as mouse and keyboard) is inadequately designed. Users are not properly supported by help functions, which, when existing, in many cases are so complex that users are further mislead. Because users have a primary interest in doing their work, they rather often tend to forget precautions and in some cases even bypass security measures when they think that their work performance is reduced.

Risks from deliberate misuse (Vulnerability Class 4)

Digital Information and Communication technologies provide many opportunities for deliberate misuse, including criminal misuse. Only few cases of criminal misuse have been reported and prosecuted. Some of these were broadly covered in the media although few caused major damage (such as the SoBig worm affecting some 100 millions of emails and several 10.000 enterprise servers). Deliberate misuse for criminal purposes has not yet significantly impacted on business and government. Consequently, both legal provisions and prosecution capabilities are less developed than in other areas of criminal law. Still, there can be no doubt that further development of ICTs will be associated with growing misuse considering that Class 1 and Class 2 vulnerabilities are so dominant in contemporary ICTs.

IMPACT OF VULNERABILITIES ON INFORMATION AND KNOWLEDGE SOCIETIES

Just as industrial technologies in the industrial age, Information and Communication Technologies will unavoidably affect many (though not all) parts of human organisations, economies, government and individual lives. As in the industrial society, ICTs are driven by supply-side concepts, with hardly any analysis of customer impact made by ICT developers. Consequently, such impacts overcome users – who seldom have the choice of avoiding ICT applications - without any possibility to understand or contain undesirable effects.

Some of this materialised in the first cycle of the Information Society (stationary operation) with complex systems where nobody can assure correctness of results in any detail. *Blind reliance* has developed among users: "this must be true because this was produced by a computer". Over-reliance and *risk acceptance* are even now, at the beginning of the 2nd cycle (network-based operation), still dominant. Just as lemmings, animals who blindly follow their forerunners, users tend to accept risks of PCs and The Internet which they feel are unavoidable. In some sense, the more technologically advanced a society is, the more risks are blindly accepted. Sociologists and philosophers observe this to be a general pattern in contemporary societies, label these as "risk societies" (Beck 1999).

Also impact on *individual attitude and behaviour* can also be observed. While email, as tool of direct and fast communication, supports personal and business needs very well, it also forces users to adapt. At best email forces users to react quickly and thus tends to dominate time management of users: if you expect a fast reaction, you must also react quickly, independent of whatever else you are doing. Replying to email overshadows any other priority. This is especially a nuisance when unwished email requires reaction, such as malicious software requesting sanitary actions or floods of undesired email (spamming).

In general, time management changes significantly when using contemporary communication systems. Similar to workers in industrial factories, users of interconnected ICT-systems behave as *slaves of engines* which they can hardly understand and control. It remains a major task of education in the Information Age to enable *users to master these engines rather than becoming controlled by them.*

ACTIONS FOR REDUCING RISKS OF CONTEMPORARY ICTS

In general, risks may be reduced in several ways:
- *"Risk avoidance"*: Information and Knowledge systems must be structured in such a way that a class of given risks cannot materialise; for example, this strategy implies that a system is designed and constructed so that it *cannot fail.*
- *"Risk reduction"*: methods and mechanisms must be realised which reduce undesired effects when some risk materialises, hopefully with

lower probability; for example, this strategy implies that a *system may fail*, but that there are curative mechanisms which *reduce the damage when the system fails*.

–   *"Risk acceptance"*: nothing preventive or curative is done ("don't care" strategy); for example, although one knows that the system is vulnerable, one simply *hopes that nothing will happen* and therefore *does nothing to prevent or reduce the risk*.

While *risk acceptance* is what the vast majority of users practice, *risk reduction* is the strategy which many enterprises and governments presently apply. To reduce the impact of crashing systems and programs, "computer viruses" and "worms", hacker attacks, mass distribution of undesired email, etc. special forms of *security software* (antivirus software, firewalls, intrusion detection systems, etc.) are *deployed to reduce threats*.

Risk acceptance may be regarded as an acceptable strategy as long as large damages can be avoided. This strategy will be *no longer acceptable, when large damages materialise* because of ineffective protection methods. By the end of the 2nd cycle of the Information Society, *interdependence of systems will have reached a degree that many small failures will combine to blackouts* similar to (though more serious than) recent power outages in the USA and Europe.

To avoid that networks become so strongly interconnected that any failure becomes "critical", the only solution available in the next decennium is to redesign basic technologies as *to become inherently safe and secure*. Regrettably mankind will only learn – as in the Industrial Age – from severe accidents, however urgent even today the need for safe and secure system designs.

BIOGRAPHY

Klaus Brunnstein is co-founder of the faculty for Informatics at Hamburg University and since 1973 professor for Application of Informatics. His special interests include: Data Protection, Computer and Network Security, Incident and Risk Analysis of IT Systems, Legal and Ethical Aspects of Informatics; Computers, Culture and Media. In 1988 he founded a Virus Test Centre (VTC) and in 1989 he started the first (2 year) course on IT Security and Incident Analysis. He is active in IFIP: the "International Federation for Information Processing" as German representative in IFIP Technical Committee-9 "Relationship between Computer and Society", as founding chairman of IFIP Working Group 9.2 "Social Accountability", and as elected IFIP President for two terms: 2002-2004 and 2004-2007. He has organised the programmes of several international conferences, among which are:

- 1989: ORAIS'89: Opportunities and Risks of Artificial Intelligence Systems, IFIP-GI Conference, Hamburg, 1989;
- 1998: Intellectual Property Rights KnowRight 2, IFIP World Computer Congress 1998, Vienna-Budapest, 1998;
- 2000: Intellectual Property Rights:KnowRight 3 & InfoEthics, Wien, September 2000;
- 2002: Human Choice and Computers, IFIP World Computer Congress2002, Montreal;
- 2003: World IT Forum (WITFOR) August 2003, Vilnius/Lithuania.

REFERENCES

Beck, Ulrich (1999) *World Risk Society*, Polity Press, Malden.

Brunnstein, Klaus (1989-2003) *Introduction into IT Security and Safety*: lectures (1989-2003) esp. addressing Incident Handling, Risk Analysis and Risk Management, Forensic Informatics.

CERT/CC (October 2002-October 2003) *Summaries CS-2003*: March 21, 2003 (A), June 3, 2003 (B), September 8, 2003 (C), November 24, 2003 (D): Source: http://www.cert.org/current/current_activity.html.

Kondratieff, Nikolai D. (1935) "The Long Waves in Economic Life," *Review of Economic Statistics* 17(6) November 1935.

Kondratiev, Nikolai D. (1984) *The Long Wave Cycle*. Richardson & Snyder, New York.

Professional deontology, self-regulation and ethics in the Information Society

Jacques Berleur
IFIP-TC9 and SIG9.2.2 Chair, Institut d'Informatique,Cellule interfacultaire de Technology Assessment (CITA, Facultés Universitaires Notre-Dame de la Paix, B. – 5000 NAMUR

jberleur@info.fundp.ac.be, *http://www.info.fundp.ac.be/~jbl*

Abstract: IFIP (International Federation for Information Processing) has been working on professional codes of deontology for the last 10 years. Lessons have been derived from that experience and most probably are also applicable in the more general field of engineering, especially when related to the Information Society. Today there is an emphasis on instruments of self-regulation in a society where it is said that legal instruments are territorial and not global. IFIP-Special Interest Group 9.2.2 has proposed a classification and has made an analysis of some of these instruments. Specific domains come forward when building a framework for regulating the Information Society. Professional deontological statements in most cases show an ethical preoccupation. The question may be raised whether this is to protect customers, clients or citizens, or whether this is for self-protection. How to implement these ethical considerations and instruments remains a challenge to be examined in the framework of the World Summit on the Information Society.

Key words: codes, deontology, IFIP, professional ethics, self-regulation

PROFESSIONAL ETHICS IN ENGINEERING AND SCIENCE

The term 'Professional deontology' is not yet fully accepted in the English terminology, but may be thought equivalent to others such as 'Code of Conduct', 'Code of Ethics', 'Guidelines', 'Standards',... Deontology refers often to the I. Kant's theory, and is used mostly in the Latin area, whereas ethics is more accepted in certain circles, mainly Anglo-Saxon, although there are also people that advocate that ethics is personal whilst professional conduct may be defined at the level of a profession: medical practitioner, lawyer, architect, engineer,... The vocabulary is far from being stable.

The relevance of ethics for engineers seems obvious *de facto*. One could be convinced by consulting the Institute of Electrical and Electronics Engineers (IEEE) 'Ethics Resources and Organizations'; or the resources, products and services of the National Institute for Engineering Ethics (NIEE). Let us select the Online Ethics Center for Engineering and Science (OECES) at CASE Western Reserve University. It seems to us a 'must' in the domain. We find there educational resources, reference material, cases, essays, and examples of Codes, maintained by the OECES itself as well as by other organisations (OECES, 2003). Interesting is also the index of topics addressed on the OECES, which reveals the kind of preoccupation that pervades the domain of activities and research:
– Confidentiality (trade secrets, database confidentiality,...) and other General Ethical Issues (lying, deception, negligence, accountability, corruption,...);
– Disciplines (Engineering [civil, electrical, environmental, nuclear], Computer Science,...);
– Diversity (general, race and culture, gender and sex);
– Educational Issues / Workplace Settings Issues (conflicts of interests, safety, design, pedagogy, competency, environmental issues, financial fraud, scientific fraud, gifts and payments, professionalism, copying, quality, risks, organisational and ethical responsiveness,...);
– Ethics Education (engineering, research, computer);
– Famous Cases;
– Legal (property, intellectual property, proprietary knowledge);
– Research Conduct (general, Internet, conflict of interest, mistake, publication and credit, research subject, working relationships).

We shall come back later on these issues after having discussed the Professional Ethics for Computer Professionals. The OECES site is unfortunately nearly totally US centred, as well as the others that we have mentioned, even if IEEE has international audience.

This does not mean that work is not done elsewhere: the European Ethics Network gathers more than 120 institutions throughout Europe (EEN 2003). It has a Journal, *Ethical perspectives: Journal of the European Ethics Network*, and publications. One of the most recent book is worth mentioning, *Technology and Ethics. A European Quest for Responsible Engineering*, because it is the first fully European contribution to the field of engineering ethics and the result of an intensive cooperation between ethicists and engineers from all the member countries of the European Union (Goujon & Hériard 2001). This book deals first with the personal responsibility of engineers: it examines the role of professional codes and the fact that engineers must cope with flexibility, shorter lines of decision and erosion of the boundaries between private and professional life; it deals then with the institutional level of responsibility: aspects of decision making in the context of business organisations, such as quality management, technology assessment procedures, business ethics committees, etc... Finally it deals with the impact of technology on society and culture, and the power of technology.

There is also, at the level of the European Commission, a European Group on Ethics in Science and New Technologies (EGE), which is an independent, pluralist and multidisciplinary body which advises the European Commission on ethical aspects of science and new technologies in connection with the preparation and implementation of Community legislation or policies (EGE 2003). During its first mandate, the EGE (1998-2000) provided 5 "Opinions" on subjects as diverse as 'Ethical Issues of human tissue banking', 'Ethical aspects of research involving the use of human embryo in the context of the 5th framework programme', 'Ethical issues of healthcare in the information society', 'Ethical aspects arising from doping in sport', and 'Ethical aspects of human stem cell research and use'. At a specific request of the President of the Commission, Romano Prodi, the Group has also produced a Report on the Charter on Fundamental Rights related to technological innovation: 'Citizens Rights and New Technologies : A European Challenge'.

We can conclude shortly saying that there is surely no vacuum in the literature, but rather a revival of the ethical preoccupation in the field of engineering. When I was a young student in engineering – 40 years ago! - we had classes on the social role of engineers; the ethical questions were mainly confined in the field of financial uprightness and honesty! Steps forward have surely been achieved.

PROFESSIONAL ETHICS IN COMPUTING: FIRST STEPS

Let us come now to a field which is to me more familiar: the 'Ethics of Computing'. We shall reflect first on the professional ethics of computing before enlarging the debate to the rather new phenomenon of self-regulation.

About traces of work on 'Ethics of Computing', I should remind a first one that I found in my archives, a text going back to 1966, hardly 20 years after the first computer was built: the *ACM (Association for Computing Machinery) Guidelines for Professional Conduct of Information Processing.* The British Computer Society which has played a leading role in promoting professional ethics in the Commonwealth was only chartered by the Privy Council, the Sovereign in Council, on July 31st 1984 (BCS 1984), but "it took the BCS almost ten years to meet the criteria required by the charter recognition, which covered, typically, such aspects as membership structure, educational standards governing eligibility for the membership grades, articles of association, rules and by-laws" (Sizer 1996). The October 1983 version of the BCS Code of Conduct, at my disposal, is however said as "superseding the previous issues". More and more the BCS stresses again today the qualification aspects. The last structural changes in 2001 made that the Ethics Committee now reports to the Qualifications and Standards Board, which replaces the old Professional Formation Board.

These ACM and BCS cases should emphasise that the conscience of 'professional ethics' was present since the 'birth' of the profession. What was stressed at the time (in the ACM Guidelines for instance) was specific behaviour in relation with the public, with employers and the clients, and with other professionals: competence in the profession, honesty, loyalty, meet the requirements and specifications of the clients, honour the profession, 'use his special knowledge for the advancement of human welfare',... The 1983 BCS code was stressing the quality of the professional conduct, the professional integrity, the respect of the public interest, fidelity, technical competence, impartiality. We shall come soon on the main content of the IFIP Codes. ACM Guidelines, BCS Code are clearly oriented towards "professionals".

But there is another story to remind, showing that things may be interpreted otherwise according to another approach of seeing the regulation of society. The Council of Europe (CoE 1979-82) has been working on the specific subject of 'legal problems connected with the ethics of data processing' for three years (1979-1982) without adopting a final resolution other than suggesting to explore 'legally' some more sensitive fields such as health care, social security, police, employment, payment and other related operations, etc. – what has been done later. The last report, prepared by Herbert Maisl for the Council of Europe, is dated September 14, 1982: 'Elaboration of an Analysis Framework for Rules of Different Nature related to the Management of Informatics'. It was examined during the meeting of September 27-30, 1982, and it was suggested to explore new problems, and to conclude the work which had been done, which meant abandoning the specific question of 'deontology' (CJ-PD[82]31, item 25). Before that decision, the Committee of experts 'Ethics of Data Processing' of the Council of Europe had questioned the European Data Commissioners Meeting (October 7-9, 1981). The second Report of Activities of the French Commission (CNIL) reports on it as follows: "The Council of Europe must take into account the spontaneous proliferation of rules of conduct as they appear in all the sectors where computers are present, and must suggest to the States Members new ways to explore, either legislative (new principles to set up), or supra-legislative (constitutional rules), or infra-legislative (recommendations or norms to be adopted by the control authorities, or directives coming from the Administration), or even infra-juridical (rules or codes set up by different organisations or associations)" (CNIL 1982). This statement of the Council of Europe debate introduces, without saying, the concept of multi-regulation which will be largely used later. It is also an expression of the compromise between a clear preference for a more normative and legal approach, and the place for 'infra-juridical' rules (the word in itself is interesting) but not for market laissez-faire.

Moreover, reflecting on this experience Prof. H. Maisl, who had been working for a long time on this issue for the Council of Europe, states that 'the rules of conduct have to reach, beyond the well structured body of computer scientists, the larger circle of computer users. We must shift, he says, from a deontology for informaticians to an objective deontology of informatics under the control of the law' (Maisl 1994).

THE EXAMPLE OF COMPUTER SCIENTISTS AND PROFESSIONAL SOCIETIES WITHIN IFIP

IFIP (International Federation for Information Processing) has no code of ethics or of conduct as an association. But its member Societies may have. We have reported at length on the experience of IFIP about ethics during the last 15 years (Berleur & Brunnstein 1996). We have analysed some 30 IFIP Codes coming from countries mostly influenced by the Anglo-Saxon way of thinking. The rules within these codes are most of the time rather simple: X is responsible about Y towards Z. X is most of the time a member of the professional society. Z is the employer, the client, the society as a whole, but also the profession itself. Y is what we have called a responsibility field.

The responsibility field is, of course, the most developed part within the codes. Five main categories emerge and regroup the different wordings as adopted by the different computer societies - we give within parentheses the number of codes concerned by the mentioned wording, provided that it appears in at least one third of the analysed codes:

- *Respectful general attitude* (/30): This attitude includes respect for the interests or rights of the people involved (15), respect for the prestige of the profession (11), respect for the interests or rights of the public (10), and respect for the welfare, health of the public and for the quality of life (10).
- *Personal (/institutional) qualities, such as conscientiousness, honesty and positive attitude, competence and efficiency* (/30): In practice, the terms conscientiousness and honesty are frequently encountered under the expressions acceptance of responsibility (19) and integrity (26). Moreover, appeals to respect for requirements or contracts or agreements (14) and to conscientious work (11) are also frequent.
- With regard to the expressions competence and efficiency, two other terms are very common: professional development and training (19) or limitation of work to the field of competence (18). Two others are also worth noting: general competence (13) and effectiveness or work quality (12).
- *Promotion of information privacy and data integrity* (/31): Confidentiality (22) is required by nearly all the general codes of the IFIP societies (13/15). Privacy in general (14) and respect for property rights (12) are appealed to quite often. Other topics are: no computer crime, no information piracy or misuse [7],...
- *Production and flow of information* (/31): The majority of the codes (23) requires flow of information to involved parties or people. Information to the public (16) is also insisted upon. Half the whole set of codes calls for comprehensive information (14).
- *Attitude towards regulations* (/30): Regulations do not appear as a major theme. Less than half the codes requires respect for the code (13), respect for the law (13), and respect for IT and professional standards (12). Some consider sanctions against a breach of the code [9].
- Regulation of the code itself is often taken into account outside the code, in the procedures.

First conclusions may be easily derived from this reminded analysis. We think that they could be similar for all the professional codes. The statements remain very general and fuzzy, and most of the time not specific to information and communication technology. On the five main responsibility fields, two are more or less directly linked to computing. And that is not yet quite sure: confidentiality was already considered in the topics enlightened by the Online Ethics Centre for Engineering and Science. General ethics, respectful general attitude, competence, quality, efficiency, accountability, fraud and property rights, as well.

The main recommendations of the IFIP Ethics Task Force to the IFIP member Societies, after its survey and analysis, were to develop more computer-related issues, to explore the most sensitive ones, to take into account the technical developments, to meet the challenges as mentioned by experts, to anticipate threats and dangers in specialised fields, to include the suggestions of international organisations (for instance the Council of Europe and its reflection on computer crime, or on privacy issues; the European Commission specialised groups),... (Berleur & Brunnstein 1996; 257-268).

Other recommendations were made also about certain weaknesses of such professional codes, in the sanction levels, the disciplinary procedures, their updating process, the status of the different computer societies, and their membership structures.

BEYOND PROFESSIONAL ETHICS: SELF-REGULATION DOCUMENTS

As Herbert Maisl suggested after the Council of Europe experience on the 'Ethics of Data Processing', we have "to shift from a deontology for informaticians to an objective deontology of informatics under the control of the law."

What has happened? Clearly, in the field of 'ethics of computing', the nineties have been marked by the proliferation, outside of the professional societies and out of the control of the law, of different kinds of codes, guidelines, standards,...: 'The Ten Commandments of Computer Ethics' (CEI 1992), 'One planet, One Net: Principles for the Internet Era (CPSR 1997), The User Guidelines and Netiquette (Rinaldi 1999), The Wartburg Online Magna Charta (Magna Charta 1997). Curiously between 1997 and 1999, all the European associations of Internet Services Providers, regrouped in the EuroISPA, wrote their own Code/Guidelines of Conduct (EuroISPA). Other Codes came into light, covering more sectoral domains, such as the health sector, the publishing sector, eCommerce, software publishers, telemarketing,... IFIP-SIG9.2.2 has established an inventory and an analysis of such 40 documents (Berleur et al. 2000). The most developed analysis has been presented during the 17th IFIP World Computer Congress, in Montreal (Berleur & Ewbank 2002). We here mention some of the most significant conclusions related to our purpose.

Content

The 'general principles' codes remain rather fuzzy, but they have a deontological content, sometimes in very normative terms: 'commandments'. They may include also consideration of correctness like the rules of Netiquette. They invite to the respect for others and for the work of the others. They could then be considered as belonging to the tradition of professional ethics or fair conduct. But they are also hardly enforceable, since there is no organisational structure and no procedure to promote and 'enact' them. They could be said 'inspirational'.

The study of the European ISPA codes shows 4 series of items at least which are recurrent in their preoccupation (for 8 codes). A recent European Commission project has confirmed our analysis (COCON 2002):

- 8 times: preoccupation about the "illegal material" (child pornography, racism propaganda,...), the necessity of youth protection especially against those who exploit their credulity, their commitment to cooperate with hotlines; but they also stress their incapacity of monitoring or controlling all their content;
- 7 times: data protection, confidentiality, email secrecy;
- 4 times: decency, no violence, no hatred, no cruelty, no incitement to commit crimes, no dissemination of propaganda material for unconstitutional organisations, respect for and care of human dignity, no ethnical, religious discrimination or on the basis of handicap or of expressed ideas;
- 3 times: fair trading, act decently with the customers, give them clear information, correct pricing information, etc.; one of them adding that they commit themselves not to promote illegal commerce!

About the sectoral codes, one could expect more specific clauses, since self-regulation is supposed to anticipate or supplement what cannot be included in the law. What clearly emerges is a more contractual character and the insistence on practices in accordance with the standards of the profession. Clauses of deontology become 'terms of services'. The Code of Ethics of the Internet Health Care Coalition, for instance, requires candour and trustworthiness, quality of information, products, services, the best *commercial* practices (our emphasis), the highest standards by Health Care Professionals (IHCC 2000). In the world of the online publishing, the French Charter reaffirms the usual rules of the profession; but they also go into details such as the number of paragraphs that can be quoted under fair dealing provisions, and without being accused of plagiarism, or specifies the rules related to links that may be created to online content (for example, links are authorised without condition provided the link opens a new window of the browser),... (Edition 2000)

In the domain of eCommerce, the 'Model Code of Conduct for Electronic Commerce' proposed by the 'Electronic Commerce Platform Nederland' (ECP-NL) offers typical clauses of good commercial practice: reliability of the information which is provided, reliability of systems and organisations, reliability of types of electronic signatures, transparency in the communication, confidentiality, respect for the intellectual property,... This code has now disappeared from the ECP-NL website!

The European Commission *eConfidence Forum*, started early 2000, seems also to be dormant (eConfidence 2001). It had produced documents such as 'General principles for generic codes of practice for the sale of goods and services to consumers on the Internet', 'Specific Guidelines for the interpretation of the general Principles', 'Guiding principles for 'approval and monitoring' bodies', and 'Options for 'Approval and Monitoring''.

The Global Business Dialogue on Electronic Commerce (GBDe) is quite more active: it holds one 'summit' each year, since 1999. Each summit provides recommendations. The last ones (Tokyo, 2001; Brussels 2002) were about consumer confidence, convergence, cybersecurity, digital bridges, eGovernment, intellectual property rights, Internet payments, taxation, and Trade/WTO (GBDe 2002). GBDe has also a Cyber Ethics Statement which is about questions linked to the diffusion of unethical material, like child pornography, anti-Semitic and xenophobic content, while fully protecting rights to free speech and expression, as well as artistic and journalistic freedom (wording of the GBDe Cyber Ethics statement) (Cyberethics 2001)! The Brussels 2002 Recommendations have now added a new chapter stressing what was until now a dissociated statement: 'Combating Harmful Internet Content.'

Comments

When looking so at more sectoral self-regulating instruments, it does become obvious that ethical concern is less apparent in the specialised work. What is becoming more prominent in this matter is contractual clauses. One example is given, for instance, in the "code of ethics" of the Internet Health Care Coalition (IHCC), which stresses, as we said, a clause about the best commercial practices.

But again, as in the case of professional codes, the clauses are fuzzy, and cannot reach the real target of protecting anybody against abuse. Their reading leaves more and more the impression that they are enacted by organisations for escaping legal liability, or for guaranteeing their own protection in case of legal proceedings.

In my opinion, the problem is multifaceted, but must be envisaged at the beginning from a double point of view: the content of the self-regulatory clauses and their development. The first question is: what are the specific issues which must be covered or not? What are the most acute ones, and from which point of view? What are the missing ones? And why? Lists of issues are numerous: we can refer, for instance to the PFIR list (People for Internet Responsibility): this list is probably too extensive to be manageable, but is interesting for consultation (PFIR 2001). The April 2003 issue of the ISOC Newsletter suggested: Cyber-fraud, Digital divide, On-line privacy, IT security, Intellectual property rights in cyberspace, Competition policy in the Internet, telecom, and IT sectors, Spam and on-line pornography, Spectrum policy for wireless Internet services. Recently, IFIP-TC9, in the Vilnius Declaration, insisted on the following issues: "Among the social and ethical concerns we strongly suggest a focus on professional ethics; access to content and technology for all; education, literacy and public awareness; multilingualism, cultural concerns; influence of globalisation; regulation, self-regulation, governance and democratic participation; intellectual property rights; specific digital policies such as eHealth, eWork, eGovernment, etc.; privacy; protection of human and civil rights; protection of the individual against surveillance; development of the quality of life and well-being; combating social exclusion; computer crime, cyber-attacks and security; employment and participative design at work; risk and vulnerabilities" (WITFOR 2003). The second question is related to the development of the issues. Most of the time, they are addressed in very short statements, where it is rather difficult to realise what is the real and mutual commitment of the parties. The statements are minimal; if not 'minimalist'. They are far from backing up or anticipating the law, as suggested by H. Maisl in his first paper to the Council of Europe (CoE 1979-82).

Then come other questions related to the legitimacy of the self-regulation instruments, and their efficiency. When speaking about their legitimacy, we cannot escape questions such as: Who is enacting, and how? Who are the authors, and how are the people concerned by the rules consulted? Is it during the process any participation of users, consumers, 'usees' (affected people)? What is the transparency of the elaboration procedures? How do the norms conform with the content of norms of higher rank? Where are the norms received and by whom? The question of legitimacy must be scrutinised with the highest care: it becomes the condition for the rules to be accepted. Unfortunately, too much is left to the discretion of the authors who enact the rules without offering any transparency, or even clarity, particularly to the people concerned and the general public. It appears that participation in making the norms is low, if not non-existent, and that actors impose on society norms which are discussed nowhere.

The question of efficiency is more related to the enforcement, the complaint procedures, and the dispute resolution. Most of the time, something is said about enforcement, but it is not easy to see how it functions: it is suggested to remove the service to a client who contravenes the rule, or to suspend the membership to a provider, for instance. But to enforce rules at an international level remains absolutely uncertain. Regarding the complaint mechanisms, they are mentioned very often, but they are not really developed, and the body who receives the complaint is rarely independent from the authors who have enacted the rules. Sanctions are also vaguely foreseen. The idea of "Alternative Dispute Resolution" (ADR) and of Cybermagistrates is going its way (ECODIR 2001). Systems of labelisation, such as BBBOnline, or WebTrust raise also new questions: methods for removing labels and the actual security of the labels appear to be in some doubt; there is no information available about how satisfactory the procedures are, what monitoring of the effectiveness of the measures there is, etc. (Gobert & Salaün 1999).

The slogans of self-regulation, 'The least State possible is the best', 'Let us avoid a greater degree of statutory regulation', or 'Let business self-regulate the Net' belong to the "knee-jerk antigovernment rhetoric of our past", and cannot persist without damage for a democratic society (Lessig 1999). They are societal lies.

... AND ETHICS?

Ethics is recognised as helping people to take more conscious and responsible decisions and act accordingly. The first question is thus: are professional deontology and self-regulation adequate instruments for that purpose?

The distinction must be made between what is related to the profession on the one side, and to society on the other. The profession has most of the time boundaries which can be more or less delineated, society not. This has deep consequences on the ethical content of the norms, due to cultural, social, legal differences, but also on the way to conceive the enforcement, the suing procedures,... The transfer of the rules from the first one to the other will be most often undue.

The functions of codes in professional deontology have been relatively well studied. We already reminded most of them in another publication, concluding: "It appears clearly that there are functions which are oriented for the sake of the profession itself - the adherence to it, its identity, the assurance of competence, the way to regulate internal conflicts, ... But there are also functions which define the boundaries between the profession and the society: they allow the public to have a look at it, or the society to know what is happening in it; they act as an appeal for responsibility at different levels - firms, society, etc." (Berleur 1996). What happens when the boundaries of the profession are unclear, as it is precisely the case for the computing profession? Again, the answer is not obvious, and depends, for instance, upon the strength of the computer societies in a country or a region. The case of the profession does not seem to us insurmountable. Ethics can find its place because member participation and public democracy are possible.

The situation of self-regulation is not the same. One could say that when it becomes more sectoral, its status is not quite different from the profession. But we have seen that when we are leaving the general principles of the deontological approach for more specific principles we are led to contractual terms, and ethics is more and more absent. We have also said that our deep conviction after having read and analysed those documents is that they are more self-protecting than self-regulating, which is the opposite of giving the public the capacity of knowing and evaluating the way our societies are facing their future. The relationship between self-regulation and the law must thus be clarified. And as professional deontology was functioning in the double role of supplementing and anticipating the legal regulation, we may wonder if the law has not to fix the framework for self-regulation. At least the representative democracy could interfere at one moment in a process which later seems more and more escape to its control. The State has not to rule everything. But it must define as clearly as possible, and in a transparent manner, the principles and the values that it wants to be respected.

In both cases, there is also a need to go further into more specific issues related to specialised fields. In other words, professional deontology as well as self-regulation must become more and more early warning. 'Specialists' are the only ones to have the capacity of such technology watch; they are in the responsibility of anticipating the future, not only in its technological dimension, but also in the social and ethical ones. But they cannot deal with that without the full participation of all people involved in the different processes. Ethics and democracy are merging together, today more than ever.

We could say that there is a future for professional deontology provided at least that it deepens the computer-related and information society issues, whilst there will be no future for self-regulation unless it clarifies democratic and participative approaches, and accepts a legal regulatory framework.

TO SUM UP... WITHOUT CONCLUDING

In my opinion, we should:
– Request more professionalism from professional bodies, i.e. clearer statements on issues in specialised fields where they develop their competence;
– Anticipate threats and dangers;
– Increase international exchange between professional societies and institutional groups, respecting the cultural, social, and legal differences;
– Reflect on the "shift from deontology for informaticians to a deontology of informatics under the control of the law";
– Question self-regulation in terms of improvement of commitment and responsibility of organisations. Is it not too minimalist?
– Increase self-regulation legitimacy by promoting large participation of all the concerned parties;
– Refrain from slogans of the past, such as "Let business self-regulate the Net" which are at risk of damaging societal fabric, and which are not favouring the cooperation between private and public;
– Clarify the relationship between deontology, self-regulation, the law, and ethics.

What Information Society needs is 'early warning', anticipating the different dimensions – including ethics – of social life. Information Society is not a technological concept, but first a social choice: it must have its norms, supported especially by ethical values.

REFERENCES

(BCS 1984): The BCS Royal Charter: http://www1.bcs.org.uk/homepages/512/ On the Privy Council and the Chartered Bodies, see : http://www.privy-council.org.uk

(Berleur 1996): Jacques Berleur, Final Remarks, in : (Berleur & Brunnstein 1996), pp. 244-245.

(Berleur & Brunnstein 1996): Jacques Berleur & Klaus Brunnstein, eds., *Ethics of Computing: Codes, Spaces for Discussion and Law,* A Handbook prepared by the IFIP Ethics Task Group, op. cit.. The background history is to be found in: Jacques Berleur and Marie d'Udekem-Gevers, Codes of Ethics/Conduct for Computer Societies : The Experience of IFIP, in (Goujon & Hériard 2001; pp. 327-350).

(Berleur et al. 2000): Jacques Berleur, Penny Duquenoy, Marie d'Udekem-Gevers, Tanguy Ewbank de Wespin, Matt Jones and Diane Whitehouse, *Self-Regulation Instruments - Classification - A Preliminary Inventory,* (HCC-5, Geneva 1998; SIG9.2.2 January 2000; SIG9.2.2 June 2000; IFIP-WCC-SEC2000), © IFIP-SIG9.2.2 http://www.info.fundp.ac.be/~jbl/IFIP/sig922/selfreg.html

(Berleur & Ewbank 2002): Jacques Berleur and Tanguy Ewbank de Wespin, Self-regulation: Content, Legitimacy and Efficiency - Governance and Ethics, in *Human Choice and Computers, Issues of Choice and Quality of Life in the Information Society,* Klaus Brunnstein & Jacques Berleur, eds., Proceedings of the IFIP-HCC6 Conference, 17th World Computer Congress, Montreal, August 2002, Kluwer Academic Publ., 2002, pp. 89-108.

(CEI 1992): The Computer Ethics Institute (CEI), Washington, D.C., The Ten Commandments of Computer Ethics, 1992 http://www.brook.edu/its/cei/cei_hp.htm

(CNIL 1982): Commission Nationale de l'Informatique et des Libertés, *Deuxième Rapport d'activités, 1er octobre 1980 - 15 octobre 1981,* Paris, La Documentation française, 1982, p.158.

(COCON 2002): Centre for socio-legal studies, University of Oxford, IAP Codes, COCON, Codes of Conduct, http://selfregulation.info/cocon/index.htm

(CoE 1979-82): Council of Europe, Committee of experts on Ethics of Data Processing, Herbert Maisl, Legal Problems Connected with the Ethics of Data Processing, Study for the Council of Europe (CJ-PD[79]8, Strasbourg, August 29, 1979; Report of Working Party n° 3 on the Ethics of Data Processing (CJ-PD[80]1); Summary of a Draft Code of Conduct prepared for the Netherlands Society for Data Processing (CJ-PD-GT3[80]1); Ethics of Data Processing (CJ-PD-GT3[81]1); The Ethics of Data Processing (CJ-PD[81]5); Ethics of Data Processing, Categories and Roles in the field of Data Processing (CJ-PD-GT3[81]2 revised); Secretariat Memorandum (CJ-PD[81]8), and the last report: Herbert Maisl, September 14, 1982: 'Elaboration of an Analysis Framework for Rules of Different Nature related to the Management of Informatics' (CJ-PD[82]19), with the Minutes of the Meeting (CJ-PD[82]31).

(CPSR 1997): CPSR (Computer Professionals for Social Responsibility), One planet, One Net: Principles for the Internet Era, 1997, http://www.cpsr.org/program/nii/onenet.html

(Cyberethics 2001): Global Business Dialogue on Electronic Commerce, Statement of Principles on Cyber Ethics, Tokyo, September 14, 2001, pp. 30-32.

(ECODIR 2001): Electronic Consumer Dispute Resolution, http://www.ecodir.org

(eConfidence 2001): European eConfidence Forum, http://econfidence.jrc.it/

(Edition 2000): Charte de l'édition électronique (Le Monde, Libération, ZDNet, La Tribune, Investir, Les Echos, L'Agefi, France) : rights and duties of the consumers, editorial content, copyright and intellectual property rights, http://www.liberation.fr/licence/charte.html

(EEN 2003): European Ethics Network: A Forum for Ethics in Europe, http://www.kuleuven.ac.be/een/Contents/introduction.html

(EGE 2003): The European Group on Ethics in Science and New Technologies, http://europa.eu.int/comm/european_group_ethics/index_en.htm

(EuroISPA): European Internet Services Providers Association, http://www.euroispa.org

(GBDe 2002): Global Business Dialogue on Electronic Commerce, Brussels Recommendations, October 29, 2002, http://www.gbde.org

(Gobert & Salaün 1999): Didier Gobert et Anne Salaün, La labellisation des sites Web : inventaire des initiatives existantes, in : *Communications & Stratégies*, 1999, n° 35, pp. 229 - 251. Better Business Bureau Inc., BBBOnLine, Code of Online Business Practices, Draft 1999, http://www.bbbonline.org/code/code.asp WebTrust Certification Services for E-Commerce Web Sites, http://www.webtrust.net/

(Goujon & Hériard 2001): Goujon Philippe, Hériard Dubreuil Bertrand, eds. *Technology and Ethics, A European Quest for Responsible Engineering*, European Ethics Network, Peeters, Leuven, 2001.

(IHCC 2000): eHealth Ethics Initiative, eHealth Code of Ethics, http://www.ihealthcoalition.org/ethics/ethics.html

(Lessig 1999): Lawrence Lessig, *Code and other Laws of Cyberspace*, Basic Books, New York 1999.

(Magna Charta 1997): Online Magna Charta, Charta of Freedom for Information and Communication, 'The Wartburg Charta', 1997, http://sem.lipsia.de/charta/gb/chartagb.htm

(Maisl 1994): Herbert MAISL, Conseil de l'Europe, protection des données personnelles et déontologie, in: *Journal de Réflexion sur l'Informatique*, no. 31, Namur (Belgique), Août 1994.

(OECES 2003): Online Ethics Center for Engineering and Science (OECES) at CASE Western Reserve University, http://www.onlineethics.org

(PFIR 2001): People for Internet Responsibility, Issues, Version of July 4, 2001, http://www.pfir.org/issues

(Rinaldi 1999): Arlene H. Rinaldi, The Net: User Guidelines and Netiquette, 1998, http://www.fau.edu/netiquette/net/

(Sizer 1996) Richard Sizer, A Brief History of Professionalism and its Relevance to IFIP, in : Jacques Berleur & Klaus Brunnstein, eds., *Ethics of Computing: Codes, Spaces for Discussion and Law*, A Handbook prepared by the IFIP Ethics Task Group, London: Chapman & Hall, 1996, ISBN 0-412-72620-3 (now available at Kluwer Academic Publishers, Boston), pp. 56-60.

(WITFOR 2003): World Information Technology Forum, The Vilnius Declaration, August 29, 2003, http://www.witfor.lt

Developments in the fields of Software Engineering
Professionalism, standards and best practice

J. Barrie Thompson
School of Computing and Technology, Informatics Centre, University of Sunderland, St Peter's Campus, St Peter's Way, Sunderland, Tyne and Wear, SR6 0DD, United Kingdom, Tel (44)191 5152769, Fax (44) 191 5152781.

barrie.thompson@sunderland.ac.uk

Abstract: Software Engineering needs to be seen as a professional discipline and for this to occur there needs to be both educational and professional infrastructures which reflect a true "engineering" ethos. A summary of recent movements in the fields of Software Engineering education and professionalism is given, followed by a more in-depth analysis of four particularly significant projects/activities: the project concerned with the Software Engineering Code of Ethics and Professional Practice, the Guide to the Software Engineering Body of Knowledge (SWEBOK) project, the Production of the Software Engineering volume as part of the CC2001 effort, and the work associated with the International Federation for Information Processing's proposals regarding the Harmonization of Professional Standards in information technology and how this relates to Software Engineering. Finally, conclusions are presented along with details of further work that needs to be undertaken.

Key words: curriculum, Higher Education, professionalism, Software Engineering, standards

THE KEY ROLE OF SOFTWARE ENGINEERING

In a world that is more and more dependent on Information Society Technologies there will be an expanding key role for Software Engineers. As stated in the documentation relating to priority thematic areas for research within the European Framework 6 initiative (FP6 ud): "Information society technologies (IST) are transforming the economy and society. Not only are they creating new ways of working and new types of business, but provide solutions to major societal challenges such as healthcare, environment, safety, mobility and have far reaching implications on our everyday life. The IST sector is now of one of the most important of the economy, with an annual turnover of EUR 2000 billion, providing employment for more than 12 million people in Europe."

Within the documentation the following needs are also highlighted:
– Technologies for trust and security;
– Addressing societal challenges;
– The development of a trusted knowledge based society;
– New computational models;
– The development of new technologies for software and systems that address composability, scalability, reliability and robustness.

All this indicates that there is a clear recognition that our future will depend on IST and that IST will have a major effect on citizens, business, and organisations. However, this recognition needs to be taken a stage further – there needs to be a realisation that such a future will depend on those who develop and deliver the systems based on IST and key amongst these will be Software Engineers. It must also be recognised that there continues to be major problems associated with many software projects. Too many projects are late, over budget and unpredictable. Often entire projects fail before ever delivering an application or, if completed, fail to deliver the required functionality. Also, in many instances when a product is delivered, the quality falls well below what should be expected from leading edge technologies. What is very clear from case study literature is that, whether one is concerned with product or process, a third vital ingredient is people. It is thus people rather than product or process that should be regarded as fundamental to the production of quality software. Other engineering disciplines, which may be considered analogous to that which supports software, have one prime feature with regard to staffing - that at the appropriate levels they have professional staff who are formally licensed or accredited as competent within their discipline. This is simply not the case within the software industry.

It is clear that Software Engineers will play a vital role in a world more and more dependent on IST. However, it is also clear that Software Engineering needs to be seen as a professional discipline - in line with other traditional branches of engineering. For this to occur there needs to be both educational and professional infrastructures and standards which reflect a true "engineering" ethos.

INTERNATIONAL MOVEMENTS TOWARDS SOFTWARE ENGINEERING PROFESSIONALISM

To support Software Engineering as a professional discipline there needs to be both educational and professional infrastructures along with appropriate standards that reflect a true "engineering" ethos. During recent years there has been some progress and clear indications that these areas are receiving higher priority. For instance:

- In 1999 and early 2000 a significant number of academic papers promoting areas related to Software Engineering professionalism started to appear in major computing journals. For example, much of the November/December 1999 issue of IEEE-CS Software and the May 2000 issue of IEEE-CS Computer were devoted to this.
- The IEEE Computer Society and the Association for Computer Machinery in 1998 created the IEEE-CS/ACM Software Engineering Coordinating Committee (SWECC) which was made responsible for co-ordinating, sponsoring and fostering all the various activities regarding Software Engineering within the IEEE-CS and ACM's sphere of operation. This committee then progressed various projects to advance Software Engineering in areas such as standards of practice and ethics, body of knowledge, curriculum guidelines, and exam guidelines. However, in summer of 2000 this formal co-operation came to an end.
- The Texas Board of Professional Engineers in June 1998 enacted rules that recognised Software Engineering as a distinct engineering discipline plus legislation that enabled Professional Engineering licenses to be issued to software engineers in Texas. However, subsequent progress has been slow. The number of licences issued has been low and each has depended on examination wavers. Also, such a system of licensing has not yet been adopted by any other USA state.
- A joint task force on Software Engineering Ethics and Professional Practice (SEEPP), established by SWECC, has developed a Software Engineering Code of Ethics and Professional Practice. The code has been

accepted by both the IEEE-CS and the ACM. In addition, many other national professional bodies for computing have reacted positively to it.
- The Guide to the Software Engineering Body of Knowledge (SWEBOK) project, which was also initiated by SWECC, has aimed at achieving a consensus view by the Software Engineering community on a core body of knowledge (BoK) for professionals within the Software Engineering discipline. The trial version of the guide that has resulted from the project has been actively promoted for public use and comment since early 2001.
- In autumn 1998, the Association for Computing Machinery (ACM) and the Institute for Electrical and Electronic Engineers Computer Society (IEEE-CS) established a joint task force to undertake a project devoted to producing a new version of their curriculum guidelines for undergraduate programs in computing. The project was named Computing Curricula 2001 (CC2001) and the task force were directed to:
  "To review the Joint ACM and IEEE/CS Computing Curricula 1991 and develop a revised and enhanced version for the year 2001 that will match the latest developments of computing technologies in the past decade and endure through the next decade."
- The early work within the curriculum project resulted in the task force deciding to divide the new Computing Curriculum 2001 (C2001) report into several volumes, each of which would focus on a particular computing discipline. One of these volumes will specifically address Software Engineering and it is expected after public review and feedback that this volume will finalised early in 2004.
- During the 1990s the International Federation for Information Processing (IFIP) were progressing a project on the Harmonisation and Acceptance of International Standards for IT Professionals, as it was believed that there was a need for rationalisation in this area. Also, the Word Trade Organisation was promoting the view that, in an era of international treaties which promoted free trade and the free movement of workers from one country to another, the establishment of standards for the qualifications of professionals. In 1998 a draft standard entitled "Harmonisation of Professional Standards" was produced. This document set out the standards of tertiary education, experience or practice, ethics, and continuing education that a customer might expect from a practitioner offering services to the public. It was hoped that eventually the document could be developed to become a Standard in the sense of ISO. During the last three years a significant number of international activities have been undertaken to promote and evaluate IFIP's harmonisation document with regards Software Engineering.
- IEEE-CS has introduced a scheme to enable it to offer a Certified Software Development Professional designation to successful applicants.

The scheme has clearly been designed to support the society's efforts to establish Software Engineering profession. The professional certification being offered by IEEE-CS has three components: exam based testing demonstrating mastery of a Body of Knowledge, extensive experience base in the performance of the work or profession being certified, and continuing professional education. Details of the scheme are available on the web and also on a CD (CSDP, ud). The latter also contains reprints of related academic articles plus copies of Software Engineering Code of Ethics and Professional Practice and the SWEBOK Guide (Trial Version).

For an international community the most significant of the above probably are:
1. The project concerned with the Software Engineering Code of Ethics and Professional Practice,
2. Guide to the Software Engineering Body of Knowledge (SWEBOK) project,
3. The Production of the Software Engineering Volume as part of the CC2001 effort,
4. IFIP's proposals within its harmonisation document and their relevance to the Software Engineering profession.
Each of these will now be considered in turn.

SOFTWARE ENGINEERING CODE OF ETHICS AND PROFESSIONAL PRACTICE

The IEEE-CS/ACM Co-ordinating Committee (SWECC) has been responsible for the creation of a joint task force on Software Engineering Ethics and Professional Practice (SEEPP). This task force, under the chairmanship of Donald Gotterbarn of East Tennessee State University, has developed the Software Engineering Code of Ethics and Professional Practice. The code is available (SEEPP ud) in two forms: a short version which summarises aspirations at a high level of abstraction and a full version which includes additional clauses. The latter provide examples and details of how the aspirations of the code should change the way persons behave as Software Engineering professionals. Table 1. details the eight principles laid out in the short code.

The code has been accepted by both the IEEE-CS and the ACM, and other national professional bodies for computing have reacted positively to it. It has received publicity both at conferences and in international journals. Also there is currently more recognition that ethical issues are becoming more and more important within computing. Nevertheless there have been some problems. The whole area of professionalism tends to be neglected by major parts of the software industry, which appears to be mesmerised with technological aspects to the detriment of human related areas and issues. There are also problems on how ethics education should be incorporated into the curricula and how it should be taught. Many believe that it needs to be fully integrated into all subjects and that positive guidance to ethical behaviour need to be in place in current software engineering courses starting in the early stages and running through the entire program.

Table 1. Principles in Software Engineering Code of Ethics and Professional Practice (SEEPP ud)

Area	Principles
1. PUBLIC	PUBLIC - Software engineers shall act consistently with the public interest.
2. CLIENT AND EMPLOYER	Software engineers shall act in a manner that is in the best interests of their client and employer consistent with the public interest.
3. PRODUCT	Software engineers shall ensure that their products and related modifications meet the highest professional standards possible.
4. JUDGMENT	Software engineers shall maintain integrity and independence in their professional judgement.
5. MANAGEMENT	Software engineering managers and leaders shall subscribe to and promote an ethical approach to the management of software development and maintenance.
6. PROFESSION	Software engineers shall advance the integrity and reputation of the profession consistent with the public interest.
7. COLLEAGUES	Software engineers shall be fair to and supportive of their colleagues.
8. SELF	Software engineers shall participate in lifelong learning regarding the practice of their profession and shall promote an ethical approach to the practice of the profession.

THE GUIDE TO THE SOFTWARE ENGINEERING BODY OF KNOWLEDGE PROJECT

The SWEBOK project, which resulted from the co-operation between the IEEE-CS and ACM on Software Engineering, is aimed at achieving a consensus view by the Software Engineering community on a core body of knowledge (BoK) for professionals in the Software Engineering discipline. The project is being run from the University of Quebec in Montreal and it is taking a three-phased approach similar to that adopted for the development of the Ada programming language, consisting of Straw Man, Stone Man, and Iron Man phases. Full details of the project and its development can be found on the project's web site (SWEBOK ud). The results of the Straw Man phase were published in September 1998 and defined the project's strategy and rationale. During April 2000 a more or less finalised edition of the Stone Man Version of the Guide (version 0.7) was released on the project's web site. This was then further refined and renamed as the Trial Version (version 0.9) which was then promoted for public use (the name was amended so that users would not incorrectly assume that the contents have been "set in stone"). In April 2001 the Industrial Advisory Board for the project approved field trials for the Guide for a two-year period. During this time the members of the project team have been very active in promoting the Guide which has been made available both as a hard copy international publication and a free download from the project site. A request for formal reviews on the use and content of the Trail Version of the guide was announced in May 2003. Once these have been analysed the results will then feed into the third stage of the project.

The SWEBOK project represents a very systematic piece of work that has attempted a broad and international approach in its reviewing process. Although the production timescales are rather long it should finally produce an authoritative and accepted BoK for the discipline. Of particular note is that the whole of the reviewing process has been visible and is available on the project's web site. Although the BoK is not yet in its final form the work accomplished so far be can be used to support the teaching within the discipline. The Guide certainly is a very useful resource for both staff and students at undergraduate and graduate levels. Also, the reviews themselves can be used to demonstrate to students the disparate views that exist across the community. However, these very positive features also have a downside. Clear difficulties with SWEBOK arise because of it being an extremely large and complex project with a rather long timetable. It is also likely that it suffered from insufficient publicity on an international level during the earlier stages, though that does appear to have improved over the last four years, where there has been a clear effort to publicise it at relevant international conferences and in journals. In addition, a close inspection of the contents of the trial version of the guide shows what could be regarded as a USA, or at least a North American continental, bias within parts of it.

SOFTWARE ENGINEERING VOLUME OF CC2001

To progress the work on the Software Engineering volume for undergraduate curricula (known as CC2001 Software Engineering or simply CCSE) there is a joint ACM and IEEE-CS initiative supported by the four groups of volunteers detailed in table 2. Also, a CCSE web site (CCSE, ud) has been established to support the project and document progress. The CCSE documents have been posted there as and when the various working groups have produced them.

Table 2. Volunteer groups supporting CCSE

Group	Responsibilities
1. Steering Committee	The organisation and co-ordination of the development of CCSE.
2. Advisory Board	Advice and links with external bodies.
3. Education Knowledge Area Group	Defining and documenting a Software Engineering body of knowledge which is known as the Software Engineering Education Knowledge (SEEK). This body of knowledge is seen as being specific to the development of undergraduate Software Engineering curricula.
4. Pedagogy Focus Group	Using the SEEK in the development of undergraduate Software Engineering curricula and the definition of undergraduate Software Engineering courses/programmes. It will also consider and advise on appropriate Software Engineering pedagogy.

In the development of the volume, extensive use has been made of external reviewing by both the general Software Engineering community and by international experts in the field. Also, a number of open events have been held at international conferences to assist in the development of the volume. Notable of these have been:

1. A workshop which was held on 25th February 2002 at the Fifteenth Conference on Software Engineering Education and Training (CSEE&T 2002), in Covington, Kentucky. (Thompson and Edwards, 2002).

2. An International Summit on Software Engineering Education (SSEE) which was held on Tuesday 21st May 2002 and was co-located with the 24th IEEE-CS/ACM International Conference on Software Engineering (ICSE2002), in Orlando, Florida. (Thompson et al, 2004).
3. A second International Summit on Software Engineering Education (SSEE II) which was held on Monday 5th May 2003 and was co-located with the 25th IEEE-CS/ACM Inter-national Conference on Software Engineering (ICSE2003), in Portland, Oregon (Thompson and Edwards, 2003).

The project is naturally large, complex, and has an expected life of 10 years or more. Progress at times has been slow at times. However, a first public draft of the Software Engineering Volume was released for public review this August via the public web site. Since then the steering committee have been actively soliciting reviews and it is hoped that the Volume should be finalised by the end of the year for publication early in 2004. In the current draft the main chapters cover:

- Guiding principles;
- The Software Engineering Discipline;
- Overview of Software Engineering Education Knowledge (SEEK);
- Guidelines for Software Engineering Curriculum Design and Delivery;
- Courses and Course Sequences.

A clear reason why the development of CCSE has been somewhat slow is that across North American Universities there are still relatively few Software Engineering programmes in comparison to the surfeit of Computer Science programs. This is simply not the case in other countries. For example, in the UK in 2002 there were at least 50 higher education institutions offering Software Engineering as a single subject degree (Edwards et al. 2003). The lack of a wide base of existing programs in North America has tended to make the developers of parts of the volume somewhat cautious in their proposals and there has a clear wish to do much from "first principles" rather than take a more "engineering" approach of "improving what already exists". In the development of the volume there have also been issues regarding what actually constitutes undergraduate education. One major issue that became very clear during the SSEE summit in Orlando was that there were two very differing viewpoints with regard to this level of education. One viewpoint was that such education is primarily about knowledge and the other was that understanding and ability should be regarded more highly.

With an expected lifetime for the volume of some 10 years there will need to be mechanisms in place that will ensure, as far as is possible, that any developed curricula remain reasonably current. Also there will be ongoing issues about addressing different modes of learning and delivery as in many parts of the world, within the next 10 years, traditional classroom delivery may no longer be the norm.

IFIP PROFESSIONAL STANDARDS INITIATIVE

In 1998 a working party within Technical Committee 3 (TC3) of IFIP produced a draft document concerned with the "Harmonisation of Professional Standards" (a copy of which can be found with the record of meetings on IFIP's site and within the publications listed later in this section). The draft standard was presented in August 1999 at the overall TC3 committee meeting in Irvine, USA and at the TC3 WG3.4 seminar held in Baltimore, USA. In the draft, introductory sections explain the overall purpose of the work, why professional standards are needed, to whom the standard will apply and clarifications concerning the terminology used. The main part of the standard then addresses the following areas, all of which are obviously relevant to Software Engineering:
- Ethics of professional practice,
- Established body of knowledge,
- Education and training,
- Professional experience,
- Best practice and proven methodologies and
- Maintenance of competence.

The IFIP harmonisation document does represent a very high level view of what is needed, nevertheless, it is very sensitive to the many complex issues that exist in the area of professionalism. It thus can provide a useful framework for further work regarding Software Engineering professionalism also it has positive attributes in that it was developed with international use as a major goal. Over the last three years conference presentations, panel sessions, participative workshops and summit meetings have been used to promote the harmonisation document within the Software Engineering arena. Of note are:
- A workshop which was held at the 2001 Conference on Software Engineering Education and Training (CSEE&T) in Charlotte, North Carolina in February 2001. (Thompson and Edwards, 2001).
- A full day workshop which was held during the 2001 International Conference on Software Engineering (ICSE) in Toronto in May 2001. (Thompson, 2001).

- A full day summit which was co-located with the 2002 International Conference on Software Engineering (ICSE) in Orlando in May 2002. (Thompson and Edwards, 2004).

The major conclusions from these activities have been:
- That there is a need for structures to support Software Engineering professionalism and that there is a need for harmonisation.
- That the IFIP document draft is a positive step and that it can be used as a framework or meta-model for Software Engineering professionalism. However, more work needs to be done on it at a detailed level.
- That the Software Engineering Code of Ethics and Professional Practice and the Guide to the Software Engineering Body of Knowledge (SWEBOK) can be seen to satisfy the first two areas highlighted in the IFIP document.
- That public safety is a fundamental driver towards professionalism. The view was expressed that one day something really major will go wrong and it will be the subsequent legislation produced by countries/states/provinces that will provide the impetus for professionalism.

CONCLUSIONS

Much of the work that has been done under the auspices of IEEE-CS and ACM high standard and it should be carried on at an international level. Software Engineering is a discipline that must operate at a global level. Other engineering disciplines such as Mechanical Engineering (which shaped the 19th century) or Electrical Engineering (which shaped the 20th century) to a great extent developed and operated within domains defined by nation states or, at least, continental boundaries. Software Engineering is different to these older disciplines in that it must be viewed in a wider context. Already we have situations where, for example, software can be specified in the USA, developed in India, and then used globally on the Internet. It is thus of paramount importance that the discipline is viewed at a global level rather than at just at the continental or national level. However, there often appears to be a lack of will or even interest in professionalism within a significant element of the community itself and in addition incompatibilities in academic and professional standards tend to act against any global harmonisation of the discipline.

Nevertheless, the IFIP report on the Harmonisation of Professional Standards should encourage international co-operation within all computing disciplines including Software Engineering. Also the very nature of the IFIP committees ensures that it has, and will be, considered at an international level and hence can not be judged to represent the view of only one sector or country. The harmonisation document essentially defines a framework, which should truly assist advancing Professional Standards if it is used in a sensitive and appropriate manner. If we consider the six areas addressed in the IFIP document and the ACM/IEEE-CS supported projects discussed in this paper we can see that there has been some real progress in the fields of Software Engineering education and professionalism. In particular:

- The ethics of professional practice is supported by the Software Engineering Code of Ethics and Professional Practice.
- An established body of knowledge is provided by the Guide to the Software Engineering Body of Knowledge (SWEBOK) and also by the Software Engineering Education Knowledge (SEEK) defined in the CCSE Volume.
- Education and training needs are supported by the CCSE Volume though this is primarily directed at education.

Yet, there is much that still needs to be done with regard to:
- Professional experience and training
- Best practice and proven methodologies
- Maintenance of competence

Therefore it is planned that to address issues in these areas a further workshop/summit will be held at the IFIP 2004 World Computer Congress in Toulouse during August 2004. It is hoped that such an event will enable opinions to be formed on:
- What are the key steps in a career in Software Engineering and other branches of computing?
- What education and training are really needed?
- How should professional behaviour be regulated?
- How is competence maintained and certified?

ACKNOWLEDGEMENT

Parts of this paper were developed from the my publications for the Fifth International Conference on The Social and Ethical Impacts of Information and Communication Technologies (ETHICOMP2001) held in Gdansk in June 2001, the Twenty-Sixth Annual International Computer Software and Applications Conference (compsac 2002) held in Oxford in August 2002, and the 2002 Asia-Pacific Software Engineering Conference (APSEC 2002) held in the Gold Coast in December 2002.

BIOGRAPHY

J. Barrie Thompson is currently professor in Applied Software Engineering, School of Computing and Technology of the University of Sunderland, United Kingdom. He is interested in educational, professional and ethical aspects associated with area of Software Engineering. He promotes development of innovative teaching approaches which are relevant to the needs of industry and promote technology transfer. He is a member of the Steering Committee which is overseeing a joint task force of the IEEE Computer Society and the ACM who are currently engaged in producing the Software Engineering Volume of an International Curricula for Computing. He also is vice-chair of IFIP Working Group 3.4 (Professional and Vocational Education and Training). The International Federation for Information Processing (IFIP) is a body which aims to co-ordinate the various facets of computing/information processing across the world. IFIP has an infrastructure of Technical Committees (TCs), each of which has a number of Working Groups (WGs). TC-3 is concerned with computers in education. Within this, W.G. 3.4 is a Working Group whose brief is education and training for IT professionals and advanced end-users.

REFERENCES AND SOURCES OF FURTHER INFORMATION

CCSE, *Computing Curricula Software Engineering: the Software Engineering volume of CC2001*, information is available from the project's web site: http://sites.computer.org/ccse/

CSDP, IEEE Computer Society Certified Software Development Professional, Web Site: http://computer.org/certification . Support CD: *Developing Software Engineering as a Profession*, Published by Institute of Electrical and Electronic Engineers, Inc, ISBN 0-7695-1899-0.

Edwards H. M. Leckenby B. M. and Thompson J. B. (2003), *A Census of Software Engineering Undergraduate Programmes in the UK*, Forum for Advancing Software Engineering Education (FASE), Volume 13 Number 03 (Issue 155), November 2003.

FP6, *Sixth Framework Programme of the European Community for research, technological development and demonstration activities*, The priority thematic areas of research in FP6 are detailed in Annex 1 of the documentation for Expressions of Interest and are held at: ftp://ftp.cordis.lu/pub/fp6/eoi-instruments/docs/eoi_annex1.pdf

IFIP International Federation for Information Processing information available from web site: http://www.ifip.or.at

SEEPP Project concerned with the Software Engineering Code of Ethics and Professional Practice. Details of this are available at the following Web Sites: http://computer.org/tab/seprof/code.htm , and http://computer.org/tab/sweec/SWCEPP

SWEBOK, Software Engineering Body of Knowledge Project , Web Site is at: http://www.swebok.org

Thompson J. B. (2001) *Report on the ICSE Workshop to Consider Global Aspects of Software Engineering Professionalism*, (which was held on 14th May 2001 in Toronto), ACM SIGSOFT Software Engineering Notes, November 2001.

Thompson J. B. and Edwards H. M. (2001), *Report on the Workshop on Achieving a World-wide Software Engineering Profession*, (which was held at Fourteenth Conference on Software Engineering Education and Training in Charlotte, North Carolina in February 2001), Journal of Education and Information Technologies, 6:4, December 2001, pp267-293.

Thompson J. B. and Edwards H. M. (2002), *Preliminary Report on the CSEET 2002 Workshop "Developing the Software Engineering Volume of Computing Curriculum 2001"*, (which was held on 25th February 2002 at the Fifteenth Conference on Software Engineering Education and Training (CSEE&T 2002), in Covington, Kentucky), Forum for Advancing Software Engineering Education (FASE), Volume 12 Number 03 (Issue 146), March 15, 2002.

Thompson J. B. and Edwards H. M. (2003), *Report on the 2nd International Summit on Software Engineering Education, Co-located with ICSE2003*, (which was held on Monday 5th May 2003 and was co-located with the 25th IEEE-CS/ACM Inter-national Conference on Software Engineering in Portland Oregon), accepted for publication in ACM Software Engineering Newsletter.

Thompson J. B. and Edwards H. M. (2004), *Post-Summit Proceedings International Summit on Software Engineering Professionalism (SSEP)*, (which was held on Monday May 20th 2002 and was co-located with the 24th IEEE-CS/ACM International Conference on Software Engineering (ICSE2002), in Orlando, Florida), available from Learning Development Services, University of Sunderland, ISBN: 1-8737757-39-5.

Thompson J. B. Edwards H. M. and Lethbridge T.C (2004), *Post-Summit Proceedings International Summit on Software Engineering Education (SSEE)*, (which was held on Tuesday 21st May 2002 and co-located with the 24th IEEE-CS/ACM International Conference on Software Engineering (ICSE2002), in Orlando, Florida), available from Learning Development Services, University of Sunderland, ISBN: 1-8737757-34-4.

The role of IEEE computer society in the Information Age

Willis K. King
Department of Computer Science, 593 PGH, University of Houston, 4800, Calhoun Blvd., Houston, TX 77204-3010, U.S.A.

w.king@computer.org

Abstract: The Vilnius Declaration key to success is finding a practical, affordable way to promote education. The IEEE Computer Society pioneered the use of the ICT technology to deliver continuing education to its members at an affordable cost. Through the use of the internet, our members can learn at their leisure basic elements in many popular programming languages, operating systems, data base and network systems and project management, among others. Members have free access to 100 Web-based training courses and in 2004 will have free access to 100 online reference books through our new Online Books program. They can also take a course to study elements of software engineering to help them prepare for an examination that would certify them as a qualified software engineer. The digital library provides practicing engineers and researchers key developments in the entire computing field in the last 15 years. Based on the success in the last five years, the Computer Society plans to expand this service to cover even more ground. Long-term plans include additional certification programs, such as a credential for software project management, and additional tutorials to include more advanced topics. Perhaps even in partnership with universities to facilitate the earning of academic credentials by members. The same technology and data base is accessible worldwide and can be used to help engineers and scientists in the developing countries to acquire new knowledge. With the help and cooperation of organizations such as UNESCO and other similar professional societies, we believe the IEEE Computer Society can play an important role in bridging the digital divide.

Key words: certification, Digital Divide, digital library, distance learning, Lifelong Learning

INTRODUCTION

One of the more obvious steps to take to bridge the digital divide is to educate the population in ICT. The most cost effective way to accomplish this goal is to enlarge and enrich the educators and practitioners of ICT in the countries concerned. The IEEE Computer Society (IEEE-CS) believes that the best way to provide a large group of people with continuing education is to make use of the very technology the ICT world has developed, namely the internet. We have the experience in serving our members by delivering educational information electronically worldwide. And we believe with the co-operation of large organizations such as UNESCO, IFIPS and other national or international professional societies, we can help to advance the educational goals of the Vilnius Declaration. Our proposal is to increase and improve the quantity and quality of ICT instructors in a country in a cost effective way, who in turn can teach the general population.

The IEEE Computer Society is a global, non-governmental, not-for-profit professional society based in the U.S.A. It has over 100,000 members from over 150 countries. It is an operation unit of the IEEE, which has a global membership of over 370,000. It publishes 23 journals in the computing area and sponsors over 150 technical conferences yearly on a worldwide basis. It has also over 200 local chapters, both for its regular and students members in six continents. Its mission, since its establishment over 50 years ago, has been to provide the computing professionals with the most comprehensive and impartial technical information.   Starting about 10 years ago, the leadership of IEEE-CS realized that the traditional way of serving its members would soon no longer be adequate. People demand information instantaneously and ubiquitously. With the rapid development in the ICT area, even experienced practitioners require constant update of their knowledge in the field they practice. In other words, the IEEE-CS needs to provide its members distance education to satisfy their demand for lifelong learning. The only way that we could deliver our service to our customers effectively is to use the very technology our profession develops viz. digitally and using the internet. Starting in 1995, we offer our publication electronically with the Digital Library. Then we provide our members an e-mail alias to facilitate electronic communication. We built an experimental virtual community on the web on distributed systems, which has since evolved to be an electronic periodical with refereed papers and open forums for over a dozen topic areas, and other community services. Since last year, we have been offering a variety of online lifelong learning services, ranging from beginning programming courses to accessing archival papers for topics in the entire information technology field. We have demonstrated that we can serve a large population of geographically dispersed customers at a reasonable cost. We believe the same system can be used to reach the rest of the world, where there is already a rudimentary infrastructure in internet technology.

THE DISTANCE LEARNING CAMPUS

To keep up with global competition and fast-changing technology in an Information Age economy, enterprises must quickly and effectively train employees while keeping costs within reasonable limits. e-Learning's popularity is rapidly growing as more enterprises see the benefits of an e-Learning solution — e.g., lower costs, rapid learning content deployment, and scheduling flexibility.

The IEEE Computer Society pioneered the use of the ICT technology to deliver on-line continuing education to its members at an affordable cost. Through the use of the Internet, IEEE-CS members can learn at their leisure basic elements in many popular programming languages, operating systems, database and network systems, and project management, among others. At the foundation of the IEEE-CS distance learning program is the following:

– *Real-world relevance* — provide content that is relevant to members' jobs, balance conceptual and contextual support with realistic scenarios and technical simulations.
– *Learning by doing* — provide a learning experience where students actively participate in the learning process, learn by doing, and are given the best opportunity to retain and apply the new knowledge after they learn.
– *Best-of-class content* — provide the best content available. The content is licensed by industry leaders including Cisco Systems, Microsoft and EMC Corporation.
– *Learning through collaboration* — provide learning within a cooperative environment, where members can benefit from sharing their individual knowledge and experiences, or can ask an expert in the field.
– *Supportive learning environment* — provide students with access to hands-on labs, class tips, online chat rooms, discussion groups, 24x7 mentoring support, and relevant career resources.
– *Self-directed* — provide the students a tool to control the learning environment, learning at their own pace.

The IEEE-CS offers the member benefit through *KnowledgeNet®* and will soon enter the third year of this valued service. Program participation exceeds 40,000 members, since the program was introduced in February

2002. Members may take any of the 100 Web-based training courses that cover a variety of computing programs and subjects including Java, XML, Windows, network security, database system management, Unix, and Visual C++ to name a few. These distance-learning courses provide learners with voice, 3-D graphics, flash animation, on-screen text, and "visual sentences" that turn complex concepts into easy-to-understand images. Access to the courses is available 24 hours a day. A minimum 56K Internet connection is required. The Computer Society is monitoring the usage of these courses and every six months, new courses will be introduced to replace some of the existing ones to guaranty that the most current and popular topics are included in the list.

The CS Distance Learning is gaining momentum as a program on several fronts. Based on this success, the IEEE-CS is looking for further opportunities to blend its current intellectual property, products and services with an expanded list of course offerings to satisfy the needs of its members. Tutorials that cover topics at a more advanced level are being looked at. There is even the possibility of partnering with universities to offer academic credentials.

SOFTWARE PROFESSIONAL CERTIFICATION

The three concepts in credentialing are often confused by individuals and professionals. Certification is not licensure. Licensure is not accreditation. Accreditation is not certification.

Certification is an occupational designation issued by an organization that provides confirmation of an individual's qualifications in a specified profession or occupational specialty. Certification is voluntary. That means it is neither a barrier nor a gate to entering or exiting a job. Certification implies an assurance than an individual possesses a specific knowledge or skill level pertaining to an occupation.

Accreditation is a designation that an organization or business has met a combination of standards and abilities that are put in place for public safety, welfare and confidence. Colleges are accredited. Hospitals are accredited. People are not accredited. The confusion stemming from accreditation may happen because the accreditation process may require certified or licensed professions to be part of the organization's operations.

Licensure is the most restrictive form of professional and occupational regulation. Under licensure laws, it is illegal for a person to practice a profession without first meeting state standards. Because states are different, so too can be the license and what it means, from state to state.

CSDP Certification

Occupational certification can be broadly grouped into three areas:
1. Certifications granted by organizations or professional associations, such as the IEEE Computer Society.
2. Industry or product-related certifications, e.g. Novell Certified Engineer.
3. Certifications granted by government agencies that train for specific jobs then validate that learning and ability to perform that job can be demonstrated.

More than 200 new IT certifications have been introduced since 1997. The number of technical certifications available to IT professionals has exploded from nearly 200 to 400 in only 3 years. More than a million of these credentials have been earned. Standards will ensure quality of preparation and testing procedures, plus help identify what each certification is intended to certify. For many, a product-specific certification, such as Novell, Microsoft or Linux certifications address needs for very specific knowledge and job requirements.

The Computer Society saw the need to cut across all products and drill down to the body of knowledge that gives definition to the constellation of computer-related certification and began developing a certification program for software engineering practitioners in June 1999. Previous to that, the IEEE-CS spent several years in partnership with industrial sponsors, other professional societies and entities, collected a commonly recognized body of knowledge in software engineering (SWEBOK). At the same time, through its work in the Standards area, the Computer Society helped establish many industrial standards in software engineering. With inputs from the industry and academia, a committee of the IEEE Computer Society identified the portion of SWEBOK and standards that a practicing software engineer should master before he/she can be viewed as aqualified software professional. It specified 11 areas in software engineering and designed a test to verify the knowledge of the candidates.

The Certified Software Development Professional (CSDP) program is a professional certification for mid-level software professionals. Software engineering professionals who see that product cycles can be as short as six months and that vendor-specific certifications will require continual training and updating maybe among the first to see the value of the CSDP's standards-based certification. Professionals who have exposure or experience to different segments that are included in the entire body of knowledge that defines the software engineering profession and now wish to put the pieces together and tie everything together with a completeness and interdisciplinary advantage will find the certification invaluable.

Generally, benefits for becoming certified may come in the form of the following.
– Additional knowledge and skills that allow you to move into a new area or perform your current job more effectively;
– Exposure to the latest software, equipment or other knowledge-based advantages you might not otherwise have;
– Increased level of expertise;
– Contact and networking with top-performing professionals in your field, around the world;
– Customer confidence based on your evidence of qualifications and suitability for the task at hand or project put out for bids;
– Added confidence.

Tutorial for Software Engineering

Complementing the launching of the certification program the IEEE Computer Society also offers a new distance-learning course. The Software Engineering Overview provides software professionals with a comprehensive review of essential software engineering principles. It also serves as a preparation course for candidates of the CSDP examination. This interactive course offers 10 modules that cover all 11 knowledge areas of the CSDP examination. Key concepts and principles of software engineering are presented with clear and concise explanations. Plus each module provides a list of references to aid in further study for each area. After completion of the course, course participants should be able to identify and understand the need for software engineering, understand software engineering principles, terms and techniques and their application to software engineering project. It also helps candidates of certification to identify areas for further review and study for the CSDP examination.

COMPUTER SOCIETY DIGITAL LIBRARY

The Computer Society library subscription package –electronic version (CSLSP-e) provides practicing engineers and researchers key developments in the entire computing field in the last 15 years. Included in the package is the entire collection of papers of the 23 periodicals published by the IEEE-CS from 1988 to present, and over 1200 conference proceedings (including most published by the society from 1995 forward). Since its launch in 1995, this digital library has steadily gained popularity among individuals and institutions alike. By now practically all major academic and industrial research libraries in the computing field are subscribers to this collection. Subscribers with access to the internet can consult the library anytime of the day and at any place in the world. We have been working diligently to enhance the accessibility of its contents by improving the search capability of the users including full text search over the 100,000 articles and papers.

Other Reference Materials

In addition to the materials published by IEEE-CS itself, beginning next year, we will offer free access to 100 online reference books to its members. Powered by Books24x7®, the Online Bookshelf will be a unique collection of 100 business and technical books on topics including Java, .NET, project management, UML, Internet security, C++ and more. Unlimited online access to this collection is available free to any member of the society. The IEEE-CS will also offer unlimited online access to collections of 500 and 2,400 business and technical books to members for a small subscription fee. As with the case of distance learning courses, this list of 100 freely access book will be changed periodically as users demand.

CONCLUSION

As we see from the above discussion, the IEEE Computer Society has successfully built an infrastructure to enable its members accessing information electronically at a reasonable cost. The information ranges from beginning level tutorials to the most advanced research papers in the field of computing. It allows the users to learn at their leisure and pace most current topics in programming languages, operation systems and issues in software engineering.

There are many aspects in solving the problem of the digital divide. Some require social engineering, as in the case between genders or generations. Others may need infrastructure improvement, as perhaps in the case between urban and rural areas. All require enormous amount of resources, both human and material. It is obviously beyond the capability of any single organization to solve the problem alone. However, in the case of training or upgrading a group of educators who have already some rudimentary knowledge in accessing the web and using the internet, the IEEE-CS can provide practical and affordable solution. Being a not-for-profit organization, our membership dues are already low. If a professional whose annual income is below certain preset level, we offer further substantial discount. With the aid and cooperation of organizations such as

UNESCO, IFIPS, and other national professional societies, we believe the IEEE Computer Society can play an important role in bridging the digital divide.

BIOGRAPHY

Willis K. King received the Dipl-Ing degree from the Technische Hochschule Darmstadt, Germany (1963), and a PhD from the University of Pennsylvania, U.S.A. (1969), both in electrical engineering. He is a senior member of the IEEE and a member of the ACM and Sigma Xi.

King has been a professor in the Department of Computer Science at the University of Houston since 1969 and served as its chair from 1979 to 1992. He worked at the IBM Laboratorien, Germany, in 1963 and 1964.

Willis King has been an active volunteer in IEEE Computer Society for more than 30 years. He was the president of the society in 2002 and serves currently as the past president. He served as vice president for area activities from 1987 to 1988 and as vice president for educational activities in 1997 and 1998. In 1999 and 2000, he was elected respectively the second and the first vice president and chaired the conferences and tutorials board of the IEEE Computer Society. He was elected and served as president-elect in 2001. Other volunteer positions he held included the local chapter chair from 1970-1975; the general chair of the second International Symposium on Computer Architecture in 1975, the IEEE Computer Society Southwestern Regional chair from 1976 to 1982, and the chair of the Distinguished Visitor's Program from 1980 to 1986. As the vice president in educational activities, he launched the model curricula project, which generated the Curricula 2001 report.

An active volunteer in computer science accreditation activities since the early 1980s, he served as an officer of the Computing Sciences Accreditation Board (CSAB) from 1985 to 1997, including as its president from 1993 to 1995.

King received the Outstanding Contribution Award from the IEEE Computer Society in 1988, the Distinguish Service Award from the Computing Science Accreditation Board in 1991 and 1993, and the Meritorious Achievement Award from the Educational Activities Board of the IEEE in 1994. In 2003, he was elected fellow of CSAB.

REFERENCES

IEEE Software Engineering Standards Collection: 2003 Edition, IEEE Press 2003

Pierre Bourque and Robert Dupuis (Editors) *Guide to the Software Engineering Body of Knowledge SWEBOK*, CS Press Dec. 2001.

Managing ICT skills profiles

Anneke Hacquebard, Steven Dijkxhoorn & Anita Erkelens
Consultancy and Research Bureau for Informatics and Education, Hummelo, The Netherlands

anneke.hacquebard@grip-project.nl; info@grip-project.nl; www.grip-project.nl

Abstract: Human skills and knowledge will be the main economic resources in a Knowledge Society. Therefore parties involved have a need to exchange information about quality and content of knowledge and skills resources. The development of a protocol for communication about skills and knowledge requires a strong international commitment, cooperation and effort on research and development, e.g. on text analysis, modelling, structuring and unified classification schemes for main domains of knowledge. As an example, ICT professionals from all over the world have identified the need for a generic description of ICT skills and knowledge. Because of the great variety in definitions, terminology and conceptual constructs, ambiguity and misunderstandings occur while trying to connect national and international educational and professional descriptions of ICT skills and knowledge. As far as we have observed, nobody has succeeded in defining such a global standard. However, it does not seem to be possible to develop one generic standard for ICT skills and knowledge either on a national scale, on the scale of an international region or on a global scale. This is based on the notion that every description of ICT skills and knowledge has its own professional, social and cultural background and stands in its own rights. A new approach is needed, such as the development and application of a referential framework and tools. A recent initiative is the development of GRIP (Generic Referential ICT Profiles), a method presented as a "common language" approach, to be used to characterise, compare and analyse existing ICT skills profiles. GRIP can be used for different purposes and with different levels of detail. This approach, originally meant for ICT profiles, seems to have a broad range of application in other domains.

Key words: ICT profile, ICT referential framework, ICT skills, ICT standards

BUCHAREST DECLARATION

On November 9, 2002, in preparation for the World Summit on the Information Society in Geneva, the Pan European Regional Ministerial Conference launched the "Bucharest declaration" on the Information Society (The Bucharest Declaration 2002). This declaration is based on seven principles and covers four themes. At least two of the themes are closely related to the management of ICT skills:

- E-Business: More Competitiveness and Better Jobs
- E-learning and E-Education: Empowering people.

> **E-Business: More Competitiveness and Better Jobs**
>
> Enterprises, both large and small, can use ICTs to foster innovation, realise gains in productivity, reduce transaction costs and benefit from network externalities. In support of this process, Governments need to stimulate, through the adoption of an enabling environment services, regulatory framework for the promotion of private investment applications and content, based on a widely available broadband infrastructure, and foster public - private partnerships. Use of digital technologies can **enhance** the role of enterprises in promoting entrepreneurship, **the accumulation of knowledge, the upgrading of skills**, and thereby increasing productivity, incomes and jobs and promoting qualitative improvement of working life. Special attention should be given to small and medium enterprises both as beneficiaries and promoters of e-business.
>
> Figure 1. Bucharest declaration

With respect to both themes we here focus on the enhancement of the accumulation of knowledge, the upgrading of skills and access to knowledge as an essential tool in economic, cultural and social development. People's knowledge and skills, often referred to as human capital, are of great importance for economy and culture. In the Information Society, or Knowledge Society, skills and knowledge are in continuous demand by business, governments, formal and non formal education and, last but not least, by individuals themselves.

> **E-learning and E-Education: Empowering people**
> E-learning is about development of skills to access knowledge, which addresses numerous issues such as local content, multi-lingual and cultural diversity and intellectual property rights. **Access to knowledge is an essential tool in economic, cultural and social development.** The potential exists for all those still outside the reach of the formal educational system to be offered education and information tailored to their need and culture. Education empowers people to overcome poverty, therefore e-learning is one of the most important issues in the bridging of the digital divide.
>
> Figure 2. Bucharest Declaration

All parties involved have to manage phenomena like:

- Constantly keeping record of people's acquired skills and knowledge;
- Frequent changes in the workforce because of a greater mobility of people and business;

– Encouragement of people, employees and employers to enhance skills and knowledge.

Domains / Professionalism		ICT knowledge and skills	Other domain knowledge and skills	Personal quality	Personal attitude
"Non-I-professionals using ready made I-technology or I-applications in their work."	ICT-user	Practical	Any All levels		
"Non-I-professionals applying I-knowledge and I-skills in areas different from informatics"	ICT-applier	Perception	Any Specialist, Conceptual	Depending on the working environment	
"I-professionals working in the field of informatics (note that the field of informatics is broad with diffuse boundaries with other disciplines)".	ICT-worker	Specialist Conceptual	Any Perception		

Table 1. Skills and knowledge matrix

Because of the large number of people involved, the kaleidoscope of educational and practical backgrounds, the variety in data about peoples' skills and knowledge, the dynamics in the development of skills and knowledge, the large amount of data, the managing of skills and knowledge is a rather complex matter. Also complexity and ambiguity of educational levels and of descriptions of skills and knowledge cause language confusion. Communication and development of understanding demand a "common language" to enable the exchange of information.

ICT SKILLS

The term 'skills' is used as a generic name to describe a person's ability to fulfil tasks and to perform roles in a particular context, usually the working environment.

Skills are part of a more generic description of a person's abilities in terms of competencies. A competency is a description of a person's knowledge, skills, qualities and attitudes that are necessary to fulfil a task in a working environment. As working environments and professionals show great differences, skills should be categorised by level of professionalism and domain (Mulder, F. & T.J. van Weert 2000). See Table 1.

Levels of professionalism are related to educational levels and levels of informal learning. Managing ICT skills in an organisation, requires a decision about the cell or cells of the "skills and knowledge matrix" to be described, to be understood or to be discussed. The matrix also facilitates a generic approach for detailing a set of profiles; in this case the matrix is used as criterion for categorisation of skills.

> A profile is a pattern of demands and expectations a person has to meet to be able to function in a working environment.
>
> Figure 3. Profile definition

Skills and knowledge descriptions concerning specific tasks or jobs are usually combined in job profiles. Educational profiles explain the skills and knowledge of graduates.

Let us take 'managing ICT' skills as an example and look at this example from a personal, an educational and a company/organisational perspective.

Personal perspective
– keeping up with professional development
– working on personal quality and personal attitude.
– to keep track of personal development, portfolios may be used.
  Prerequisites for this are that the portfolio has already been created during higher or vocational education and that the person is able to maintain the portfolio during his or her working life.

Educational perspective
– flexible assessment procedures, taking into account student quality and competency, acquired by formal and informal learning;
– facilities to build student's portfolios;
– getting used to the idea that the portfolio belongs to the student;
– facilities for the maintenance and long term availability of portfolios.
  This requires a long term vision, a guarantee that portfolio sites, portfolio databases or some other information system are long lasting. The mobility of the workforce and the scale of the data to be maintained demand international cooperation to find solutions so that lifelong availability can be guaranteed.

Company and organisational perspective
– Implementation of competence management systems
– This is a way to manage competencies of personnel, to get to know for what tasks and roles they are employable, what their educational needs

are, in order to follow career paths. For the people involved competence management systems offer a clear view on their capabilities, on how to handle ambitions and on planning for reaching personal goals and objectives.

To meet the goals stated in the Bucharest declaration and other declarations there is a need to manage ICT skills and knowledge on the level of persons, education, companies and organisations; this implies more precise definitions of the specific skills and knowledge required.

PROTOCOL FOR COMMUNICATION

To be able to manage ICT skills and knowledge, parties involved should be able to understand and overview the respective skills profiles and job profiles or, in other words, the quality of persons. Attempts to make quick and simple overviews of, for example, the ten leading profile documents, deadlock in differences of approaches.

ICT professionals from all over the world have been asking for a generic description of ICT skills and knowledge. For instance, during the Working Conference Meeting Global IT Skills Needs - the Role of Professionalism, 35 specialists from 14 countries, discussed the need of a standard or referential standard (IFIP OECD WITSA Joint Working Conference Meeting Global IT Skills Needs – the Role of Professionalism October 25th – 27th 2002). One of the conclusions of the conference was: " develop a high level reference model covering groups including IT professionals, IT practitioners and others, to identify the differences in obligations associated with different types of work; to assist the closer matching of employer requirements with educational provision".

Management of skills and knowledge is not only of interest to the ICT domain. There is an urgency to deal with the complexity of knowledge and skills because of its spin-off to industry, education and society. As far as we can see, nobody has been able to define an accepted (global) standard for ICT skills. Experience tells that it is not possible to develop one generic standard for ICT skills and knowledge, either on a national, international or a global scale. Because of the great variety and ambiguity in definitions, terminology and conceptual constructs misunderstandings and deadlocks occur while trying to construct sound overviews of various descriptions of ICT skills and knowledge.

REFERENTIAL FRAMEWORK

Communication about the content of ICT skills and about knowledge needs some "vehicle": a framework or a common language approach.

It seems attractive to use an existing set of profiles as a standard or as a framework, but experience shows that one profile model cannot serve "all" existing profiles because of differences in levels, goals, structure and terminology.

A profile model is a coherent structure, serving as a basis for the description of a set of profiles.

Figure 4. Profile model definition

Looking at the skills and knowledge matrix, we notice differences in the mastering of skills and knowledge (practical, perception, conceptual). It is unlikely that one of the existing profile models covers all these areas. It can be considered to use different models for different levels of mastering. This approach carries the risk that it will lead to several sub-standards, existing next to one another, and even worse, with inevitable overlap. As barriers between formal and informal learning diminish, the use of different standards may create confusion.

The choice for a particular standard usually does not meet all the local requirements, opinions and goals. Therefore the use of an existing profile model and the profiles belonging to it, as a standard for "all" profiles, will not solve the problem of diversity and ambiguity. It is better not to focus on just one model serving as a standard, but to look for another solution. And also: *existing profile models have not been not build as referential model and will not meet the requirements of a referential model.*

By using an independent referential framework - instead of one of the existing profile models - as an "umbrella" a lot of problems can be solved. Such a framework is not available; it has to be newly designed and built.

REQUIREMENTS FOR A REFERENTIAL FRAMEWORK

What functions should the framework have? A framework is an instrument to link and clarify the main issues of existing and new models describing ICT skills on issues like: content, terminology, skills levels, educational level, competency, tasks and roles, target groups, actor, etc.

Who are supposed to use the referential framework? The existing ambiguity, ordered chaos of models, is a concern of international organisations, governments, employer organisations, employers, employee organisations, educational institutions on all levels, international and national professional organisations, parties involved in formal and non formal learning etc. The great variety of users of the framework makes necessary a detailed research on the way those parties want to use it, the tools needed and to (re)define and refine the requirements.

What are the requirements for a referential framework? A framework should be transparent, not too complicated for quick use, offer more advanced (may be inevitably more complex) features for detailed research. A framework has to be clear about its own descriptive terminology, structure and operation instructions. The framework should offer a guideline for building, rebuilding and maintaining of profiles and profile models. The framework should offer a way to map profiles, in order to be able to compare mappings of profiles' content from different origin and different architecture.

What kind of tools are needed? Advanced digital tools have to be developed to support the process of analysis. A huge amount of data has to be handled, text analyses have to be performed and statistics have to be derived. To facilitate reference and distribution of results, easy to use tools for users have to be available. Please note that the development of a referential framework does not imply the development of a new profile model or a new set of profiles.

An independent referential framework:
- does not contain profile descriptions;
- offers an unified approach to:
  - analyse profile model structures and terminology used;
  - categorise profile models and profiles;
  - independently maps profiles content onto domains of knowledge and personal behaviour;
- is a basis for:
  - benchmarking or comparing profiles and profile models of different origin and different architecture;
  - maintenance and review of existing profiles and profile models;
  - a unified guideline for development of profiles and profile models.

As an advantage of a referential framework for benchmarking and comparison, existing profiles do not have to be changed. Another reason to keep existing models alive is that people and organisation usually have very good reasons for the use of "their" terminology and "their" profiles. For many good reasons they are not very willing to drop their models and profile descriptions.

In due course, as a spin-off, while the framework approach gets acknowledgment and achieves a status generally accepted, the referential framework can encourage unity of terminology and unity of structures of profile models.

GRIP: GENERIC REFERENTIAL ICT PROFILES

As a recent initiative GRIP can be mentioned. GRIP is a "common language" approach, to be used to characterise and analyse existing ICT skills profiles or as a guideline for building new profiles.

GRIP has been developed to meet the demand explained in the preceding paragraphs: how to understand and deal with differences and similarities of profile models and profiles. GRIP does not offer new profiles, it is a method meant for analysis of profile models and profiles. In GRIP the analysis of profile models and profiles is approached in two ways:

a) Analysis of a profile model
   A checklist is used to summarise the document and profile model characteristics. For instance, terminology used, structure of the document, structure of the profiles are recorded (Hacquebard, Dijkxhoorn & Erkelens 2003).

b) Analysis of a single profile
   The content of separate profiles is mapped onto classification schemes for ICT or other disciplines and personal attitude and behaviour. These mappings are "independent", in the sense that profiles are mapped onto classification schemes which do not belong to any of the profile models.

GRIP maps ICT profiles onto:
- UCSI, Unified Classification Scheme for Informatics, a detailed mapping;
- Other domains (alpha, beta, gamma), a global mapping);
- Personal attitude and behaviour, a detailed mapping.

Our experience is that instruments, like text and linguistic analysis, are needed to perform these tasks. Use of just a yellow marker in documents to identify content, has turned out to be unworkable.

The outcome of a GRIP analysis of a profile model contains:
- GRIP A: one report about the profile model as a whole;
- GRIP B: as many reports on content as there are profiles in the profile model.

A selection of profile models and profiles has to be made before the analysis is performed. Criterion can be to select profiles belonging to the same category (cell) in the "skills knowledge matrix" It depends on the objectives of the GRIP analysis, which profiles and profile models are to be included. It is obvious that a decision has to be taken about the purpose of the analysis, before applying the GRIP method to profile models and profiles.

Possible purposes are:
- evaluating and benchmarking existing profile models and its profiles;
- comparing job profiles and educational profiles;
- comparing ICT curricula of several IT departments of educational institutes in a region;
- serving as a guideline for revision of existing profiles or for building new profiles.

GRIP can be used for different purposes and in different levels of detail. This approach, originally meant for ICT profiles, seems to be also applicable to a broad range of other domains as the summary and mapping approach can be performed for any profile model or profile, given that classifications schemes for the specific domains of knowledge are available. GRIP can be seen as a first step towards a referential framework. Summaries of analysis outcomes of GRIP will be published on the GRIP site.

CHALLENGE

Developing a referential framework, that covers the needs of a great variety of users and serves as a protocol for communication about skills and knowledge, requires a strong international commitment, cooperation and effort on research and development, e.g. on text analysis, modelling, taxonomies, structuring and use of unified classification schemes for main domains of knowledge. It is a challenge for international cooperation to gather expertise and initiatives towards developing a referential framework, acknowledged by as many parties as possible, in order to enable the managing of ICT - and other - knowledge and skills profiles.

Such a protocol is a necessary condition for the realisation of the Bucharest declaration, especially for the themes "E-Business: More

Competitiveness and Better Jobs" and "E-learning and E-Education: Empowering people". A referential framework could be used as a basis and starting point for the development of tools for the management of ICT skills and knowledge in order to realise enhancement of the accumulation of knowledge, the upgrading of skills and access to knowledge. However, the principles of a referential framework are not restricted to the ICT domain, neither are problems about ambiguity of job and educational profiles. This justifies a strong international effort and a professional approach of the development of a referential framework for descriptions of skills and knowledge.

BIOGRAPHIY

Anneke Hacquebard is General Manager of a Consultancy and Research Bureau for Informatics and Education. She has been active in research on IT Skills models, for example in an initiative by IP-HOB (Informatics Platform Higher Education, Business and Trade), partially founded by the Ministry of Education, Culture and Science, the NGI and the VRI (Dutch Computer Societies). She is one of the originators of the GRIP (Generic Referential ICT Profiles) method. She is working on research on referential frameworks for descriptions of professional quality.

Steven P. Dijkxhoorn is an ICT specialist and is studying Educational Science and Arts at the University of Nijmegen. In developing GRIP, his concern has been the application of text analysis and modelling.

Anita Erkelens B.Sc. graduated from Larenstein University of Professional Education, Velp. She spent several years working on international projects performed by Arcadis Euroconsult and the FAO and was involved in research and vocational training. In developing GRIP, she contributed to the design of classification schemes and modelling.

APPENDIX

Table 2. UCSI classification, level 1 and level 2

1	Computer systems	2	Software systems	3	Information systems	4	Context of informatics
1.1	Computer architecture	2.1	Programming languages	3.1	Information bases (IB)	4.1	Management and informatics
1.2	Interfacing and peripherals	2.2	Software architecture	3.2	Information systems architecture	4.2	Domain specific and dedicated systems
1.3	Communication and networks	2.3	Software engineering (SE)	3.3	Information systems engineering	4.3	Informatics operational environment
1.4	Operating systems and system software	2.4	Artificial intelligence (AI)	3.4	Interaction and Presentation (IP)	4.4	Informatics in society
1.5	Miscellaneous	2.5	Theory of computing	3.5	Theory of information systems	4.m	Miscellaneous
1.m		2.m	Miscellaneous	3.m	Miscellaneous		

Table 3. Domains of knowledge, level 1 and level 2

A	Alpha sector	B	Beta sector	C	Gamma sector	I	Informatics sector
A1	Economics	B1	Mathematics, sciences & physics	C1	Humanities & art	I1	Informatics
A2	Jurisdiction & administration	B2	Engineering & technology	C2	Socio-cultural	I2	Business Informatics
		B3	Medical			I3	Medical Informatics
						I4	Technical Informatics
						I5	Social Informatics

Table 4. Domains of personal behavior and attitude, level 1 and level 2

D	Rational	F	Performance	H	Attitude	O	Development
Da	Analytical	Fl	Executive	Hb	Professional or vocational	Oa	Ambition
Dc	Conceptual	Ft	Team	Hp	Personal	Oe	Experience
Do	Other	Fz	Individual			Ol	Intellectual faculties(l), learning potential

REFERENCES

Hacquebard, A., S. Dijkxhoorn & A. Erkelens (2003) *GRIP: A different outlook on profiles, Overview.* Adviesbureau voor Informatica en Onderwijs Hacquebard bv, Hummelo. [http://www.grip-project.nl].

IFIP OECD WITSA Joint Working Conference *Meeting Global IT Skills Needs – the Role of Professionalism* (October 25th – 27th 2002) Summary proceedings, Woking, Surrey, United Kingdom. [http://www.globalitskills.org/proceedings.pdf]

Mulder, F. & T. J. van Weert (2000) *Informatics Curriculum Framework 2000 for higher education [ICF-2000],* UNESCO, Paris. [http://www.ifip.or.at/pdf/ICF2001.pdf]

The Bucharest Declaration (2002) *Towards an Information Society: Principles, Strategy and Priorities for Action.* Bucharest Pan-European Conference in Preparation of the World Summit on the Information Society: 9 November 2002. [http://www.wsis-romania.ro/menu/home/Documents/declaration.html]

Enabling ICT adoption in developing Knowledge Societies

Colin Harrison
Director of Strategic Innovation, IBM Global Services EMEA Strategic Outsourcing, IBM Zurich Research Laboratory, Säumerstrasse 4, 8803 Rüschlikon, Switzerland

ch@zurich.ibm.com; http://www.research.ibm.com/people/c/colinh

Abstract: The deployment of ICT in its present form requires simultaneously mastering of many skills and having a developed infrastructure of human and technical resources. These are frequently lacking in regions remote from the affluent neighbourhoods of major cities, whether in developed or developing economies. Moreover, potential users in these developing Knowledge Societies may have different needs or a different balance of needs from the established user base. Such neighbourhoods of major cities already provide an ICT ecology and their users' needs are heavily pre-determined by the prevailing Internet culture. In developing Knowledge Societies, however, the introduction of ICT – like any major infrastructure investment – is likely to be a communal decision, prioritised against other needs, and conditioned by local values. So the introduction of ICT into such a community needs to consider 1) what needs do we wish to meet, 2) what ICT infrastructure can meet those needs, and 3) how can we bootstrap the ICT ecology that will enable the deployment to become rapidly self-sustaining. The technology selection and deployment process thus requires a much broader assessment and the choices may – paradoxically – be wider than for an established Knowledge Society. In my contribution, I will propose a framework for preparing for the creation of a new Knowledge Society that is based in part on current experiments in developing economies and in part on a view of the evolution of the underlying technologies.

Key words: sustainable technology, technology diffusion

INTRODUCTION

The scope of this discussion is the deployment in developing knowledge societies of information and communication technologies (ICT) for the direct use of the society. I assume that the fundamental challenges faced by such societies are: limited personal or social funding, limited technical infrastructures (electricity, communication networks), and limited skills. But other issues need consideration: what is the priority of ICT compared to other local needs (health, education, transportation), what needs and wants of the society as a whole should be addressed by ICT, and what will be the impact on the society of deploying ICT (new job opportunities, new cultural threats)?

This paper considers the unique characteristics of ICT and from these develops possible roadmaps to deployment based on the following principles:

1. What needs and wants do we wish to address by this deployment?
2. What forms of ICT can addresses the various classes of needs and wants?
3. How do we create a sustainable ICT infrastructure that addresses those needs and wants?
4. How do we evolve such an ICT infrastructure as the needs and wants and abilities of the society evolve?

THE DIFFUSION OF TECHNOLOGIES

As William Gibson famously remarked, "The future is here. It is just not widely distributed yet." It is a characteristic of all technologies that they are invented, piloted, and perfected locally among small groups of users before beginning the long haul to universal use. Geoffrey Moore has written about this transition in "Crossing the Gap" (Moore 2002). Early adopters of a new technology face considerable challenges: poor performance, poor reliability, poor support, and (relatively) high cost. They are motivated to use the technology because it provides some high value that is not otherwise available.

Consider the initial deployment of cellular telephones in the United States during the mid- to late-1980s. Initially the coverage provided was small and patchy, and there was no provision for roaming. The services were expensive relative to fixed network telephony. They also represented an enormous, speculative investment on the part of the cellular carriers. Cellular telephony found early adopters among people who were mobile within small areas, typically a single town, and for whom the advantage of mobile telephony offset the high monthly bills – lawyers, doctors, and property agents.   Christensen has written widely on how emerging technologies find their initial markets (Christensen 1997).

An early adopter community is an essential element for the successful deployment of a new technology. Without this community there is no way to "cross the chasm" and attract the much larger mainstream community. In the absence of an early adopter community, it is tempting to turn to government to carry the technology across this secondary gap. However, governments have historically proven to be poor sponsors for emerging technologies.

For the technology to "cross the chasm", it must appeal to users with less strong motivations. Only by expanding beyond the early adopters to the much larger mainstream communities can it achieve the returns to scale that enable the developers to recover their initial investments. Reaching larger user groups in turn means that the price of the product or service can be reduced. Early cellular telephony users frequently had bills of $1,000 per month for service within their own town; today such service might cost as little as $25 per month.   Broad adoption also requires new levels of performance, reliability, ease of use, and support.   Some of these characteristics – performance, reliability – become imbedded in the device or the service as the developers capture their learning into the technology; their costs are easily recovered as the scale of use expands. Others – ease of use, support – have costs that increase in parallel with the scale of use. To succeed, the new technology must also find a financial model that enables its deployment to be sustainable.

The secondary diffusion of a technology into regions away from its origins poses a new set of challenges. While by the time of secondary diffusion, many of the core performance and reliability issues will have been resolved, it may be harder to find an early adopter community that will justify the initial deployment costs. The regions of secondary diffusion will often have different needs and wants from those of the technology's birthplace and there may be no early adopter community that can derive high value from the initial deployment.

For a technology to diffuse into a developing society there must be an infrastructure that contains the following features:

1. A minimum level of service or benefit that meets a key set of needs and wants of some members of the society.
2. A set of technologies that is affordable by those members of the society.
3. A collection of centralised and end-user skills that is sufficient to deploy, operate, and use the technology to meet the key wants and needs.
4. A set of technical, commercial, and educational mechanisms that enables the new technology to become indigenous rather than exotic and to expand its accessibility to the broader community.

CHARACTERISTICS OF ICT

ICT shares many features with other 19-20C technologies, but it takes some of these to extremes not found elsewhere. In part these account for some of its amazing success in the last 30 years, but they also represent challenges for its successful deployment in emerging knowledge societies.

Positive returns to scale

A key feature of the economics of ICT is the extreme positive returns to scale, which is due in large part to the success of Very Large-Scale Integration. Computers of the 1960s were largely hand-built using tens of thousands of discrete transistors and they cost of the order of a million US dollars. Today's microprocessors are manufactured by enormously expensive semiconductor factories with a globally-integrated supply chain and produce chips containing hundreds of millions of transistors that cost of the order of hundreds of US dollars. This massive reduction in cost per function has been achieved because of the dramatic increase in volumes – from a few thousand computers per year in the 1960s, to hundreds of millions of computers and other digital devices today.

A consequence of this positive-returns-to-scale model and of the massive investment required to develop a new device or a new manufacturing capability is that it appears unlikely that non-mainstream semiconductor or storage technologies will play a major role in deploying ICT into developing knowledge societies. While we can imagine meeting the needs and wants of such societies with simpler devices (chips), such approaches face insurmountable challenges in "crossing the chasm"; that is, in achieving sufficient volumes to reduce unit costs below the costs of mainstream devices.

Generations of technology

A second key feature of ICT is the generations of technology. These generations have been dominated to some degree over the last 20 years by the symbiosis of Intel's microprocessor generations and Microsoft's operating system generations. Through the leadership of these and other ICT vendors, the industry has had a well-defined rhythm.

So an alternative approach to deploying ICT in developing knowledge societies might be to use older generations of technology. Towards the end of a product lifecycle, the investment costs have been recovered and it becomes a "milk cow" that can be sold just above its marginal cost of production to drive volume until the industry is ready for the next generation. So it is conceivable to produce low-cost personal computers using older generation microprocessors, smaller amounts of physical memory, storage, and lower resolution displays.

However, devices are only part of the story and there is a limitation of this approach due to the vertical integration of the ICT industry as a whole. The generations of technology sweep through the industry from top to bottom: from the machines that produce the chips, to the machines that assemble devices onto boards and into systems, to the operating systems and device drivers that provide useful platforms, and the applications that provide end-user services. While there are capabilities within the industry to provide legacy support, they extend only so far back. The further we fall behind industry generations, the greater the likelihood that we have to provide this legacy support ourselves.

Centralisation of skills

A third key characteristic is the ability of ICT to embed knowledge within its products. For example, early personal computing required detailed knowledge of how to set up the signals that enabled a modem to interact with the Central Processing Unit (CPU). This involved hours of exploring combinations of "DIP switch" settings. Today, that knowledge is captured in industry standards and the operating systems that automate the integration of devices with the CPU. In the context of introducing technology into developing knowledge societies, this is good; we do not have to develop support skills for these tasks.

This is true of all technologies, for example cars, but ICT is an extreme case. In the 20C we saw that wherever a society was able to amass competency in metallurgy, mechanical engineering, chemical engineering, and manufacturing engineering, an automobile manufacturing industry could arise from the ground up. This was true also for ICT manufacturing up to circa 1970, when the degree of integration within the industry was relatively small. But since then ICT has spread by the export of knowledge from a small number of developed knowledge societies into regions with lower manufacturing costs. The transition of that exogenous knowledge into indigenous knowledge does take place, but slowly relative to the development of the knowledge. The Internet Protocols and the work of the open standards bodies in general are great exceptions to this principle.

Since ICT is a delivery mechanism for skills, it is also possible for needed skills to be provided outside the area of deployment. For example via remote system management can to some extent reduce the need for local skills. On the other hand, such knowledge is and will remain exogenous to the society, and as time goes by, local skills that have been developed will (rapidly) become obsolete. So in planning a sustainable deployment, we need to consider carefully what the profile of needed skills will be over the coming years. This planning also needs to take into account the future uses of ICT will be used within the society.

Product or service?

Today's social use of ICT reflects the values, skills, and finances of the middle classes of westernised societies. It is predicated on the ability of a family to invest of the order of US$1,000 in a personal computer that is replaced every 3-5 years and of the order of US$20 per month in communication services. The ability to master the evolving end-user role has been viewed as a key social skill, just as the ability to perform automobile servicing was once viewed as a test of manhood.

The perceived benefits of this investment of money and learning are communication (e-mail), access to information (entertainment, education, financial management), e-commerce (shopping, micro-scale business), and community (chatrooms, clubs, on-line gaming). These have evolved rapidly since 1992, when e-mail became recognised as the "killer application" that induced of the order of a few million end-users to make these investments. Each succeeding set of benefits has built on the infrastructure created by the previous set and has been driven by – or been enabled by – succeeding generations of technology.

Western middle-class society experienced ICT as a set of products with supporting services for two main reasons: 1) a significant part of the required technology needs to be installed at the point of use and 2) the rapid evolution of the applications and benefits – combined with the rapid evolution of the technology - meant that no viable business models evolved for a service-based approach to ICT. Contrast this with the telephone which emerged and has continued to be experienced primarily as a service with few private individuals owing and operating a home Private Branch Exchange (PBX).

We see in the above a number of economic and technical features that facilitate the introduction of ICT into developing knowledge societies and others that are inhibitors. In the next section we consider how these key features of ICT suggest different approaches to this deployment of ICT from those that have driven its deployment in Western middle-class societies.

A FRAMEWORK FOR THE ADOPTION OF ICT
IN DEVELOPING KNOWLEDGE SOCIETIES

These key features of ICT can give a set of policies for considering the set of questions given in the Introduction:
1. What needs and wants do we wish to address by this deployment?
2. What forms of ICT can address the various classes of needs and wants?
3. How do we create a sustainable ICT infrastructure that addresses those needs and wants?
4. How do we evolve such an ICT infrastructure as the needs and wants and abilities of the society evolve?

Needs and wants

From Moore (Moore 2002) we saw the need to find the initial community of early adopters who will provide the seed from which a broader deployment will grow. These early adopters will be people for whom (simple) ICT can provide high value relative to the costs. Some such community must be found in which ICT provides a better solution to some important want or need than existing alternatives. Table 1 shows possible assessments of these wants and needs and their possible values for various members of a developing knowledge society.

Table 1. is not prescriptive; it merely suggests possible benefits for some of the roles in societies. The people in these roles may or may not see that ICT can effectively meet their wants and needs. Even in highly developed Western countries, farmers are not early adopters of ICT, preferring instead to use the telephone. However in considering a deployment, such an analysis should be conducted. If no group or groups of people can be found to form an early adopter community of a sufficient size to provide the financing – through the purchase of goods and services - for the needed infrastructure (see below) then it is unlikely that a sustainable deployment can be achieved.

What form of ICT?

The above analysis will give a good idea of what initial set of ICT services is required. However the categories of service (e-mail, access to information, and so forth) can be provided with differing levels of service compared to those expected in developed knowledge societies. For example, in developed knowledge societies e-mail has lost its original connotation of an asynchronous service and is often used as a form of chat service with almost real-time interchanges. Some years ago the MIT Media Lab demonstrated that e-mail could in fact be provided without any networking infrastructure. The researchers created a mobile system transported on a motorcycle that could perform daily pickup and delivery of e-mail from servers located in isolated villages. Similar creativity can be applied to access to information, e-commerce, and communities.

Second, the technical level of the systems employed can be assessed. While we have noted that there are limitations to the use of older generations of technology, what compromises can be made in function and performance that will enable older technologies to be employed?

Table 1. Social roles and potential benefits of ICT services

Role	Email	Access to information	e-commerce	community
Student	Simple e-learning	Opportunity for higher levels of education	Selling/buying coaching services	Collaboration with other students. Experience of other cultures
Artisan	Simple e-commerce	Awareness of demand for styles, prices	Micro-business, access to markets	New markets, sharing of techniques
Farmer	Simple e-commerce	Awareness of demand for products, prices	Micro-business, choice of markets	Cooperatives, negotiating organisations
Nurse or general Medical Doctor	Request to hospitals, pharmacies	Treatments, medical alerts, epidemics, continuing education	Buying medical supplies, choice of suppliers	Communities of practice
Merchant	Simple e-commerce	Awareness of demand for styles, prices	Buying stock, choice of suppliers	Buying cooperatives
All	Staying in touch with extended family and friends	Awareness of the region, country, world.	Increased choice	Experience of other cultures
Alternative solutions	Mail, personal travel	Libraries, newspapers, circulating letters	Local markets, wholesalers	Local trade associations, newsletter, travel to meetings

Third, should ICT be made available as a product or a service? The Western middle-class experience is perhaps a reflection of the decline of social capital in Western societies during the last 60 years (Putnam 2000). Other societies may prefer a more communal model for the initial ICT deployment, even if it is initially sponsored by the early adopters.

How to achieve sustainability?

Inevitably the introduction of ICT into developing knowledge societies begins with exogenous technology and skills. How should the society go about achieving sustainability? There is a tendency to want to create local industries for PC manufacture, but in reality this offers poor opportunities for sustainability.

Table 2. Value network for ICT products and services

Technology	Accessibility	Opportunity for local value addition
Personal computer management	High	High, probably difficult to provide remotely
Server operation (architecture, service definition, tariffs)	High	High
Server management	High	High, though could be provided remotely
Network operation (architecture, policies, tariffs)	High, but needs high skill levels	High
Network management	Moderate, but needs high skill levels	High
Software installation	High, mainly skill-based, little capital required	High
Software development	Moderate through Open Source Software communities, but needs very high skill levels	Low, will not add significant value in the short term
Hardware repair (computer and network equipment)	Moderate, mainly skill-based, but may require some electronic test gear, also spare parts	High
Hardware installation	High, mainly skill-based, little capital required	High
Assembled systems	Very low (very high capital investment)	Minimal
Chip and storage	Extremely low (massive capital investment)	Zero

Table 2. shows a partial hierarchy of the value network for ICT with a rough assessment of the ease with which each layer can be performed within a developing knowledge society and of the potential for contributing value to that society.

Because of the key features of ICT discussed above, the potential for contributing near the base of the stack is limited and the value that could be achieved is also limited. Moving upwards, better opportunities exist in acquiring the system and network architecture and operational skills needed to deploy ICT to meet the wants and needs of the community. These skills do require a quick mind, but do not require higher education. They are among the most valuable skills in the ICT industry and so there is a risk of leakage of the people with these away from the community (see however the discussion below on How to Evolve for the Future).

The ICT industry has a well established model of levels of support for products and services. Level 1 support can address the most common problems encountered by users, say, the top 90%. Level 2 support can address more serious technical questions and perform diagnosis and some repair; this covers a further 8-9%. Finally Level 3 support has deep knowledge of the products and services and can diagnose any problem. The deployment can thus leverage this hierarchy to provide Level 1 support on site with the others accessible more rarely via telephone or e-mail.

So the sustainable deployment model is based on:

1. Acquiring externally the minimal technical systems needed to support the wants and needs of the early adopters;
2. Temporarily importing the skills to set up and begin operation of these systems;
3. Making these available either as products or as services to an early adopter community;
4. Developing locally the Level 1 support skills that must be delivered locally;
5. Relying on remote Level 2 and Level 3 support;
6. Eventually developing some Level 2 and 3 capabilities.

How to Evolve for the Future?

If a sustainable deployment of systems and skills, founded on meeting the affordable wants and needs of a community of early adopters, can be established, then there is good potential for "crossing the chasm" into broader adoption. This results in economies of scale reducing the cost of service and enabling additional services to be provided. This is predicated on a general economic expansion enabled or supported by the early adopter usage.

There are models of economic development based on the introduction of motor cycles leading to a growth in per capita income and in turn to a transition to cars; each stage generating new skills and commercial opportunities. No such bootstrap model based on ICT has been demonstrated. There have been claims of increased commercial productivity in Western societies, but these have been challenged recently (Henwood 2003).

The early history of mobile telephone operators and Internet Service Providers (ISP) is perhaps instructive. The strategy of both these groups was initially to find early adopters to enable them to get into the business and then to reduce prices and acquire as many mainstream users as possible with the aim of being bought out by a large telecommunication company or large ISP. This seems a likely course of events for these deployments of ICT into developing knowledge communities. Just as large, mainstream airlines are effectively unable to establish low-cost subsidiaries, so large telecommunication companies or ISPs are unable to engage in this form of ICT deployment. However, once a deployment is successful, it will form an attractive takeover target for such an operator and this could bring a significant capital infusion into the community.

My own belief is that the greatest value of ICT into developing knowledge communities comes not from the services themselves, but from the consequential emergence of new patterns of thinking within the community and the development of new skills that enable the community to take part in the broader economic activities of the region or worldwide.

CONCLUSIONS

The deployment of ICT in developing knowledge societies poses a set of choices that reflect the unique features of these technologies, the economics of bootstrapping new technologies, and the specific needs of the community. Alternatives exist to the Western middle-class model of ICT, notably the delivery of ICT as a set of services rather than products. The opportunities for developing sustainability in the ICT deployment are most likely to be in the operation and management of the systems and services rather than in the development of hardware or software systems and this can bring highly valuable new skills to the community. The ultimate success of such a deployment is probably identified by the acquisition of the community's systems and services into a larger organisation, but will leave the community enriched with new skills and opportunities.

BIOGRAPHY

Colin Harrison joined IBM in San Jose, California in 1979 and has held many technical leadership positions in IBM's product businesses, in IBM's Research Division, and currently in IBM's IT services business. In 2001 he established IBM's Institute for Advanced Learning. Following his university studies, he spent several years at CERN developing the SPS accelerator. He then returned to EMI Central Research Laboratories in London, and lead the development of the world's first commercial MRI system. With IBM he has enjoyed a career leading from micromagnetics to medical imaging, parallel computing, mobile networking, intelligent agents, telecommunications services, and knowledge management.

Colin studied Electrical Engineering at the Imperial College of Science and Technology and earned a PhD in Materials Science. He also studied Physics at the University of Munich. He is a Fellow of the Institution of Electrical Engineers (UK) and a Senior Member of the Institution of Electronic and Electrical Engineers (USA). He is a Chartered Engineer (C.Eng.) and a European Engineer (Eu Ing). He was a founder member of the Society of Magnetic Resonance in Medicine (USA). He is also an expert advisor to the Swiss Academy of Technical Sciences. He has been a visiting scientist at MIT, Harvard Medical School, and Lawrence Berkeley Laboratory.

Colin Harrison has been awarded 26 patents. He has published some 40 scientific and technical papers and talks and a successful book on Intelligent Agents. He is an invited speaker at European universities on the impact of information technology on the nature of work, business organisation, and industries.

REFERENCES

Moore, G. A (2002) *Crossing the Chasm*, HarperBusiness, ISBN 0-0605-1712-3.

Christensen, C. M. (1997) *The Innovator's Dilemma*, Harvard Business School Press, ISBN 0-8758-4585-1.

Putnam, R. D. (2000) *Bowling Alone: The Collapse and Revival of American Community*, Simon & Shuster, ISBN 0-684-83283-6.

Henwood, D. (2003) *After the New Economy*, New Press, ISBN 1-5658-4770-9.

Sustainable development and the Information Society
From Rio to Geneva

René Longet

President of Equiterre, Partner for sustainable development, 22 rue des Asters, 1202 Geneva, Switzerland, Tél. +41223299929

longet@equiterre.ch; www.equiterre.ch

Abstract: Our world is facing very strong challenges and needs: a) Globalisation that divides as well as it unites the world, 2) natural resources and heritage that are to be preserved for the purposes of the whole mankind and for the coming generations, and cannot be reserved to a minority and for the needs of the present time only, 3) equity in access of these resources is to be established among humans. A better world is not only possible, it is absolutely necessary. We must use our technological, scientific, social, administrative capacities to face the needs of billions of people staying behind, to manage in a sustainable way the natural resources. We all know these necessities. But why are our actions so shy, so slow, and so tiny? Sustainability gives us a goal for the introduction and the use of new technologies. Sustainability asks for public awareness, capacity building, education (and not global soap opera), free exchange of information and opinion, monitoring, ethical guidance, social justice in access to information and tools, to define limits of abuse also.

Key words: capacity building, citizenship, Digital Divide, information, sustainability, World Summit

CITIZENSHIP

Citizenship implies a need for and means to access information, but also the ability to deal with information. Information therefore is a very important issue:
– Information means knowledge;
– Information means capacity to understand what happens;
– Information means freedom from false beliefs and gives more self-confidence.

More information technology not automatically provides for more and better information. Information technology can enable ignorant people to come to awareness, can connect isolated people, can enable a society to gather better data and as a result monitor its own development. But none of these important goals will become reality without a serious effort and a strong political will. The technology alone will not achieve this by itself.

Information technology is relatively cheap, not complicated to handle and very flexible. It is therefore much more difficult for censors to combat against Internet publications, than to close down traditionally published newspapers (see the examples in: Iran. Zimbabwe, China, etc.). Information technology helps to build up civil society and citizenship, especially in the Third World.

But if information technology means no limit to communication, how can one stop the invasion of undesirable e-mails, how can one combat racist, pornographic or violent messages and eliminate paedophile web pages without eliminating the free exchange of political opinions and scientific - and other - information through the worldwide net?

Information technology can enable people to become better informed, enable governments and civil society organisations to better monitoring of developments in their country or city. But what is the message passing through the newly opened channels? By watching the global soap-opera-programs broadcasted by most TV channels, people in the Third World might think, that people in the industrialised world are all rich, lazy, own big cars and big swimming-pools, having plenty of beautiful girls around them… It encloses the people all over the world in a virtual reality that will substitute the real world and undermine the will to change it.

WHAT IS THE LESSON?

The lesson to be learned from all this is that a new technology will never be able to give us social, ethical or human progress without a good social, ethical and human framework. The real issue is not only to deal with the so-called Digital Divide, but to also assure that the information technologies will help to improve our societies and our lives. Thus, such a framework should concern also the field of environmental issues.

The environmental issues of the wide spreading of information technologies are the followings: materials (poisonous for the environment, possibly for men as well, if misused or abandoned without treatment), ability to be repaired, ability of its components to be recycled. The question is: is there a life-cycle approach in the branch? What is the amount of toxic or non toxic waste? What is the increase of the global energy consumption? What is the source of this energy, especially in the Third World?

WORLD SUMMITS

In order to achieve the goals of sustainable development, we need a better information capacity. To have a better information capacity we need to use information technology in a specific direction. This should be the goal of the World Summit on the Information Society (WSIS) now held in Geneva. To this purpose it is of great importance that this Summit creates a bridge with the other UN Summits of the 90's and of the beginning of the Century: Habitat, Social Summit, Food, Women's rights, Development, Millennium, etc. The United Nations should not separate the results of all these Summits, but unite them in a global view. This view is already in existence: it is the view of Sustainable Development, as formulated at the World Summit on Sustainable Development at Johannesburg in 2002.

Sustainable development means to face the real needs of our world. What is going on in the real world today? Globalisation divides us, but also unites us. The most important issues of our time are:
- To guarantee human rights all over the world;
- To promote peace and coexistence between different cultures;
- To implement a sustainable way of development all over the world.

All our technical means should be employed to achieve these goals. This is naturally also true for information technologies.

SUSTAINABILITY

What does sustainability aim at? The most important goals as noted in the Action Plan adopted at Johannesburg on the 4th of September 2002 are:
- Eradication of poverty;

- Changing of our ordinary patterns of production and consumption;
- Managing the natural resource basis of social and economic development.

Sustainability wants to give a simultaneous answer to issues of poverty, in-equal development and misuse of natural resources. The idea was defined in 1987 (Brundtland-Report of the UN, entitled Our Common Future), recognised internationally in Rio 1992 (UNCED) by all States, representatives of civil society, business, the UN, and reinforced in Johannesburg 2002.

We like to quote some social goals of the World Summit on Sustainable Development in Johannesburg:

- Halve by 2015 the proportion of people whose income is less than 1$ a day, who suffer from hunger, who stand without access to safe drinking water, who have no access to basic sanitation.
- Ensure that children - boys and girls - will be able to complete a full course of primary schooling.
- Increase employment opportunities taking into account the fundamental rights at work.
- By 2020 achieve a significant improvement in the lives of slum dwellers.

INITIATIVES

In the environmental field we find the commitment to promote sustainable consumption and production patterns, to support regional and national initiatives to accelerate this purpose, and to de-link economic growth and environmental degradation through improving efficiency in the use of resources and production processes. These measures are achieved through approaches like life-cycle analysis, the polluter pays principle, consumer and producer information, increase in eco-efficiency, support of cleaner production, improving of social and environmental performances of business, encouraging dialogue between enterprises and their stakeholders, using environmental impact assessment procedures.

20% of all inhabitants of the World use 45% of animal proteins, 58% of the energy, 84% of the paper, and 88% of the vehicles.
USA: 3% of the global population, 25% of the global energy consumption

---

Figure 1. The resource divide 1

In Johannesburg, following commitments were made about this issue:
- Integrate energy efficiency and accessibility into development programs;
- With a sense of urgency substantially increase the global share of renewable energy;
- Reduce greenhouse gas emissions;
- Implement transport strategies in the sense of sustainability.

---

850 millions of human beings suffer from hunger
850 millions are an-alphabets.
1 billion have an income less than 1 $ per day.
1 billion are without an acceptable housing.
1,1 billion are without access to water of drinking quality.
2,4 billions have no access to basic sanitation facilities.

Figure 2. The resource divide 2

A DUTY TO HUMAN DEVELOPMENT

We, the developed world and its technology, have the duty to assure a human development all over the world and the basic conditions of human dignity. We must develop but not in the way we do this now. Indeed, since the 19th and even more in the 20th century we are going in the wrong way:
- The oceans are over-fished and many fish stocks are depleted;
- The climate change has obviously begun;
- The threat over the non-renewable resources is real;
- The loss of biodiversity, and of the tropical forests is going on;
- Desertification is also going on;
- Pollution of atmosphere, water and soil is also a reality.

Within the last 50 years, the fossil energy consumption has been multiplied by 5 and the greenhouse gas emissions have followed the same way. But we notice extreme differences between the regions of our world: 20 to/y/person in the USA, 10 in Germany, 4 in Mexico, 1.5 in Brazil, the quantities are much smaller in Asia and Africa. The acceptable level is 1 to/y/person.

Other environmental commitments of interest taken in Johannesburg were:
- Prevent and minimise waste and maximise reuse;
- Sound management of chemicals though their life cycle;

– Protect freshwaters and introduce a sound management of the seas inclusive sustainable fisheries;
– Facilitate the protection of the ozone layer;
– Promote sustainable agriculture, land use, mining and forestry;
– Achieve by 2010 a significant reverse in the loss of biological diversity.

In many paragraphs of the Plan we find the necessity of information to assure monitoring, and good governance. The Action Plan reassures that nothing is possible without the work field of scientists, the gathering of information and the diffusion of the collected knowledge about the state of sustainability. But we have to underline that the information technologies have to be used to enforce sustainable development and to assure that they are not misused in the other way. They should help in monitoring situations and in return, in mobilising people in the direction of good solutions. This should be the goal of the present Summit.

APPENDIX

WORLD SUMMIT ON SUSTAINABLE DEVELOPMENT ACTION PLAN, QUOTATIONS ABOUT INFORMATION

16 (c) Collect and disseminate information on cost-effective examples in cleaner production, eco-efficiency and environmental management, and promote the exchange of best practices and know-how on environmentally sound technologies between public and private institutions;

20 (m) Promote education to provide information for both men and women about available energy sources and technologies;

23 (f) Encourage development of coherent and integrated information on chemicals, such as through national pollutant release and transfer registers;

25 (b) Facilitate access to public information and participation, including by women, at all levels, in support of policy and decision-making related to water resources management and project implementation;

36 (b) Establish by 2004 a regular process under the United Nations for global reporting and assessment of the state of the marine environment, including socio-economic aspects, both current and foreseeable, building on existing regional assessments;

38 (g) Promote the systematic observation of the Earth's atmosphere, land and oceans by improving monitoring stations, increasing the use of satellites, and appropriate integration of these observations to produce high-quality data that could be disseminated for the use of all countries, in particular developing countries;

47 (b) Encourage ongoing efforts by international financial and trade institutions to ensure that decision-making processes and institutional structures are open and transparent;

52 Assist developing countries and countries with economies in transition in narrowing the digital divide, creating digital opportunities and harnessing the potential of information and communication technologies for development, through technology transfer on mutually agreed terms and the provision of financial and technical support, and in this context support the World Summit on the Information Society.

110 Assist developing countries, through international cooperation, in enhancing their capacity in their efforts to address issues pertaining to environmental protection including in their formulation and implementation of policies for environmental management and protection, including through urgent actions at all levels to:

(a) Improve their use of science and technology for environmental monitoring, assessment models, accurate databases and integrated information systems;

(b) Promote and, where appropriate, improve their use of satellite technologies for quality data collection, verification and updating, and further improvement of aerial and ground-based observations, in support of their efforts to collect quality, accurate, long-term, consistent and reliable data;

112 Use information and communication technologies, where appropriate, as tools to increase the frequency of communication and the sharing of experience and knowledge, and to improve the quality of and access to information and communications technology in all countries, building on the work facilitated by the United Nations Information and Communications Technology Task Force and the efforts of other relevant international and regional forums.

114 Examine issues of global public interest through open, transparent and inclusive workshops to promote a better public understanding of such questions.

121 Integrate sustainable development into education systems at all levels of education in order to promote education as a key agent for change.

124(a) Integrate information and communications technology in school curriculum development to ensure its access by both rural and urban communities, and provide assistance particularly to developing countries, inter alias, for the establishment of an appropriate enabling environment required for such technology;

128 Ensure access, at the national level, to environmental information and judicial and administrative proceedings in environmental matters, as well as public participation in decision-making, so as to further principle 10

of the Rio Declaration on Environment and Development, taking into full account principles 5, 7 and 11 of the Declaration.

129 Strengthen national and regional information and statistical and analytical services relevant to sustainable development policies and programmes, including data disaggregated by sex, age and other factors, and encourage donors to provide financial and technical support to developing countries to enhance their capacity to formulate policies and implement programmes for sustainable development.

132 Promote the development and wider use of earth observation technologies, including satellite remote sensing, global mapping and geographic information systems, to collect quality data on environmental impacts, land use and land-use changes, including through urgent actions at all levels to:

(a) Strengthen cooperation and coordination among global observing systems and research programmes for integrated global observations, taking into account the need for building capacity and sharing of data from ground-based observations, satellite remote sensing and other sources among all countries;

(b) Develop information systems that make the sharing of valuable data possible, including the active exchange of Earth observation data;

(c) Encourage initiatives and partnerships for global mapping.

133 Support countries, particularly developing countries, in their national efforts to:

(a) Collect data that are accurate, long-term, consistent and reliable;

(b) Use satellite and remote-sensing technologies for data collection and further improvement of ground-based observations;

(c) Access, explore and use geographic information by utilizing the technologies of satellite remote sensing, satellite global positioning, mapping and geographic information

164 All countries should also promote public participation, including through measures that provide access to information regarding legislation, regulations, activities, policies and programmes. They should also foster full public participation in sustainable development policy formulation and implementation. Women should be able to participate fully and equally in policy formulation and decision-making.

BIOGRAPHY

René Longet was born in Geneva in 1951. He received the grade of "Licence en lettres" of the University of Geneva. He was engaged in education, politics and publications. Since 2001 he is President of Equiterre, a Swiss Non-Governmental Organisation with as its purpose the promotion of sustainability in society and politics. He was member of the Swiss Delegation to the World Summit on Sustainable Development held in Johannesburg in 2002. He is also member of many committees and boards, such the steering committee of the Centre for Technology Assessment at the Swiss Science and Technology Council. He is Mayor of the city of Onex nearby Geneva (17.000 inhabitants) and has contributed to many publications within the field of sustainability.

The "e-well"

ICT-enabled integrated, multisectorial development of rural areas in the least developed countries

Antoine Geissbuhler & Ousmane Ly
Medical Informatics Service, Geneva University Hospitals, Geneva, Switzerland; Mali telemedicine network coordinator, REIMICOM, Bamako, Mali

antoine.geissbuhler@hcuge.ch; oussouly@keneya.net

Abstract: ICT has the potential to improve the quality and efficiency of cooperation and development efforts. However, there is a risk that ICT-enabled development projects will worsen the digital divide between urban and rural areas. It is therefore crucial to involve rural areas early in these development efforts, to make sure that their specific needs are addressed. Experiences in the field of health and healthcare have shown the potential of multilateral co-operation, the involvement of the periphery and the importance of supporting local contents development. Obviously, one of the challenges is to find sustainable ways to promote the ICT development of remote, rural communities. It is likely that specific solutions will be needed in different cultural and social contexts, but that replicable concepts will be identified. The proposed project, the "e-well", based on a pilot implementation in rural Mali, will enable a reciprocal learning from parallel experiences of ICT-enabled multi-sectoral, integrated development plans, in several villages from different settings in Western Africa.

Key words: developing countries, Digital Divide, Internet access, rural areas, sustainable development, tele-medicine

DEVELOPMENT EFFORTS CAN WORSEN THE "DIGITAL DIVIDE"

Technological progress usually creates inequities, at least initially. This is notable in all societal aspects, at the global level, and in the information society in particular, where these inequities separate the information-rich from the information-poor populations, creating the so-called "digital divide".

In most development efforts, there is a natural tendency to develop centrifugal projects, starting from the main cities where the wealth and the expertise is often concentrated, towards the periphery, following the deployment of infrastructures such as roads and telephones. When dealing with ICT, these development efforts carry the risk of inducing additional inequities. Such "induced digital divide" could probably be prevented by giving special attention to the needs of the remote, underserved areas, in order to have them connected early to the development efforts, and therefore in a position to influence the whole project.

Global connectivity enables new forms of deployment, which can be decentralized from the beginning. What is now relatively easy at the technological level, is somewhat more complicated at the contents level. Creating contents of quality requires skills that are more difficult to solve by technology, and more difficult to decentralize. This "contents gap" between the rich and poor countries, between the urban and rural areas, is therefore an other, non-technological, expression of the digital divide.

LESSONS LEARNED FROM THE TELEMEDICINE PROJECTS IN MALI

Telemedicine tools enable the communication and sharing of medical information in electronic form, and thus facilitate access to remote expertise. A physician located far from a reference centre can thus consult colleagues remotely in order to resolve a difficult case, follow a course over the Internet, or access medical information from digital libraries or knowledge bases. These same tools can also be used to facilitate exchanges between centres of medical expertise: health institutions of a same country as well as across borders. The potential of these tools is particularly evident in developing countries where specialists are few and generally concentrated in large cities, and where geographical distances and the quality of the infrastructure often hinder the movement of physicians or patients.

A project named « Keneya Blown » (the "health hallway" in Bambara language), was initiated in 2001 by the Mali University Medical School in Bamako, and financed by the Geneva government and the Geneva University Hospitals. Several goals were set: a) develop and use Internet-based connections between the national and regional healthcare institutions, b) implement basic services such as e-mail and a medical Web portal, c) implement a low-bandwidth, Internet-based distance learning system (http://www.unige.ch/e-cours), d) evaluate the feasibility of long distance collaborations for continuing medical education and tele-consultations.

Various types of collaboration have been enabled by the project:

- North-South tele-education: topics for post-graduate continuing medical education are requested by physicians in Bamako; courses are then prepared by experts in Switzerland and then broadcasted over the Internet from Geneva. New courses are produced and broadcasted on a bi-monthly basis, on a variety of topics. The material is also saved and can be replayed from the medical Web portal. Typically, these courses are followed by 50 to 100 physicians and students in a specially-equipped auditorium in the Bamako University Hospital, and also by smaller groups or individuals in the Segou and Tombouctou regional hospitals, and in other French-speaking countries in Africa: Senegal, Mauritania, Chad, Morocco, Tunisia.

- Webcasting of scientific conferences: several sessions of international conferences have been broadcasted, with simultaneous translation in French, in order to make the presentations accessible to colleagues in Mali, where the practice of the English language is still limited. Using the instant messaging feature of the system, remote participants can intervene and ask questions to the speakers.

- South-South tele-education: post-graduate and public health courses, developed by the various health institutions in Bamako, are webcasted to regional hospitals in Mali and to other partners in Western Africa. The contents produced are anchored in local, economical, epidemiological and cultural realities, and provides directly applicable information.

- South-North tele-education: medical students training in tropical medicine in Geneva follow courses and seminars organized by experts in Mali on topics such as leprosy or iodine deficiency. The exposure to real-world problems and field experts enables a better understanding of the challenges for developing and implementing healthcare and public health projects in unfamiliar settings.

- North-South tele-consultation: the same system can be used to send high-quality images enabling the remote examination of patients or the review of radiographic images. Tele-consultations are held regularly, in areas

where expertise is not available in Mali, such as neurosurgery or oncology.
-   South-South tele-consultation: physicians in regional hospitals can request second opinions or expert advice from their colleagues in the university hospitals, via e-mail, including the exchange of images obtained using digital still cameras.
-   South-North tele-consultation: the case of a leprosy patient, followed in Geneva, has been discussed using the tele-consultation system, and enabled the expert in Bamako to adjust the treatment strategy.

This three-year experience in deploying a telemedicine network in Mali has led to several useful observations:
a) ICT may increase the digital divide;
b) ICT can reduce some of the health system's fractures;
c) local contents development is a major challenge, and
d) South-South networks are most relevant.

ICT may increase the digital divide

It is obvious that ICT serve first those who have access to it, and therefore tend to put at higher priority exchanges between technologically equipped partners. In our experience, initial telemedicine activities were oriented towards leading-edge medical practices, and the exchange of expert advice between teaching centres. Primary care or public health topics were not seen as priorities. This can be explained by the type of partners involved initially, representing mostly tertiary care, teaching medical centres.

Paradoxically, the main benefit of telemedicine should be seen in remote areas, where expertise is lacking and where the main problems are likely to be at the primary care and public health. Medical infrastructure available in such remote areas also prevents the application of leading-edge, often technologically advanced, practices.

ICT can reduce the health system's divides

At the same time, the deployment of telemedicine tools in regional hospitals greatly improved continuing medical education opportunities and enabled some forms of tele/consultations, in particular for second opinions on radiological images.

Further to the periphery of the network, the implementation of a satellite-based internet connection in a rural hospital was a strong enough incentive that motivated a physician to accept the job in the remote village, provided he could stay in contact with his colleagues and follow training courses via the e-learning environment.

Local contents development is a major challenge

Local medical content is a key for the acceptance and diffusion of health information, and is also essential for productive exchanges in a network of partners. It enables the translation of global medical knowledge to the local realities, including the integration of traditional knowledge. Medical content-management requires several levels of skills: technical skills for the creation and management of on-line material, medical librarian skills for appropriate contents organisation and validation, and specific skills related to the assessment of the quality and trustworthiness of the published information, including the adherence to codes of conducts such as the HONcode.

South-south networks are most relevant

At the content level, there is a steady demand for North-South distance learning. However, several topics for seminars, requested by physicians in Mali, could not be satisfactorily addressed by experts in Switzerland, due to major differences in diagnostic and therapeutic resources, and to discrepancies in the cultural or social contexts. For instance, there is no magnetic resonance imaging capability in Mali and the only CT-scanner has been unavailable for months. Chemotherapeutic agents are too expensive and their manipulation requires unavailable expertise. Even though diagnostic and therapeutic strategies could be adapted, practical experience is lacking, and other axes for collaboration must be found. A promising perspective is the fostering, through decentralized collaborative networks, of South-South exchanges of expertise. For example, there is neurosurgical expertise in Dakar, Senegal, which is a neighbouring country to Mali. A tele-consultation between these two countries would make sense for two reasons: a) physicians in Senegal understand the context of Mali much better than those from northern countries, and b) a patient requiring neurosurgical treatment would most likely be treated in Dakar rather than in Europe.

Based on the lessons learned during the pilot project, a larger, four-year project involving six countries of Western and Northern Africa has been launched in 2003: the RAFT project (Telemedicine Network in French-speaking Africa). The following aspects are emphasized:

- The development of a telemedicine infrastructure in teaching medical centres, and their connection to national and international computer networks, in order to foster multi-lateral medical expertise exchanges, with a predominant South-South orientation.
- The use of asynchronous, collaborative environments that enable the creation of virtual communities and the control of workflow for getting expert advice or second opinions, in a way that is compatible with the local care processes.
- The deployment of internet access points in rural areas, with the use of satellite technology, enabling not only telemedicine applications but also other tools for assisting integrated, multi-sectoral development, and, in particular, education, culture and the local economy. The mini-VSAT technology, recently deployed over Western Africa, offers an affordable, ADSL-like connectivity. Sustainable economical models, based on the successful experiences with cybercafés in Africa, are being developed to foster the appropriation of this infrastructure by rural communities.
- The development and maintenance of locally- and culturally-adapted medical contents, in order to best serve the local needs that are rarely covered by medical resources available on the internet. New tools are being developed: regionalized search engines, open source approaches, and adapted ethical codes of conduct. The Cybertheses project (http://www.cybertheses.org) and the resources from the Health On the Net Foundation (http://www.hon.ch) are used to train physicians, medical documentalists and librarians.

THE DIMMBAL.CH PROJECT: MULTI-SECTORAL INTEGRATED DEVELOPMENT IN RURAL MALI

Sustainability

One of the key challenges of development projects is their economic sustainability. Sustainability can be improved by enabling simultaneous development activities in multiple sectors (education, health, economy, culture). This requires a significant effort, geographically-focused, involving most of the stakeholders of the community, in order to reach a significant increase in development, compatible with long-term sustainability of the process and results.

Reciprocal enabling

Collaboration between a scientific research team and the local authorities of the commune of Dimmbal, in Dogon country, Mali has led to the elaboration of a four-year multi-sectoral integrated development plan (http://www.dimmbal.ch). The strategy is to obtain a critical mass of cross-enabling activities so that their appropriation by the population and long-term sustainability are improved.

For example, it is accepted that young adults leave the village during the dry season in search for employment, usually in the large cities, where they are more likely to catch sexually-transmitted diseases, and bring these back to the village. Developing local industries could therefore, by occupying these young adults, reduce potentially dangerous behaviours.

Many of these activities can be enabled by a satellite-based internet connection: the telecentre. Main sectors and activities of the projects include:

– Infrastructure: construction of a telecentre and media library, training of local coordinators to the techniques of management and ICT.
– Education: extension of the primary school buildings, reading and writing classes for adults, continuing training for teachers via the internet.
– Economy: support to the development of local enterprises and industry, valorisation of local, traditional knowledge, reforestation and plantation of medicinal herbs.
– Health: additional wells and forages, extension of the local dispensary, creation of a laboratory for medical analyses, integration of traditional medicine in the practice of the dispensary.
– Culture: valorisation of history and traditions by ethno-historical and archaeological research, support of the local theatre team, publication of paper-based and electronic documents.

The project also develops evaluation tools, in particular to measure the hypothesized enabling of the various activities and to validate the economical sustainability of the approach.

THE "E-WELL" PROJECT: LEARNING FROM PARALLEL DEVELOPMENT EFFORTS

The "e-well" symbolizes the central, multi-functional role of ICT, as an enabler of development efforts. It is likely that success factors and obstacles in such projects will be educative to similar projects in other settings. The goal of the « e-well » project is to run several development projects in different rural settings in developing countries, and to evaluate, compare and share results in order to learn collectively from the various experiments. This would enable a south-south learning network focusing on knowledge about the technical, social and economical engineering of rural development projects. The expected outcome is a better understanding of the potential, success factors, impact and sustainability, of integrated, multi-sectoral approaches to the development of rural areas in different settings.

The "e-well" project plans to include 6 different sites with four-year development plans, and various coordination, evaluation and sharing activities between the local coordinators of each site. The project is designed to run over 7 years (2004-2010), for a total budget of €6'500'000, under the coordination of the AGENTIS, a UNITAR project, part of the decentralized cooperation programme, dedicated to exploit the potential of information and communication technologies for development and social initiatives.

BIOGRAPHY

Antoine Geissbuhler is a Professor of Medical Informatics at Geneva University School of Medicine, and Director of the Division of the Medical Informatics at Geneva University Hospitals.

A Philips European Young Scientist first award laureate, he graduated from Geneva University School of Medicine in 1991 and received his doctorate for work on tri-dimensional reconstruction of positron emission tomography images. He then trained in internal medicine at Geneva University Hospitals under the direction of Prof. Francis Waldvogel. After a post-doctoral fellowship in medical informatics at the University of Pittsburgh and Vanderbilt University, he became associate professor of biomedical informatics and vice-chairman of the Division of Biomedical Informatics at Vanderbilt University Medical Center, under the mentorship of Prof. Randolph Miller and Prof. William Stead, working primarily on the development of clinical information systems and knowledge-management tools. In 1999, he returned to Geneva to head the Division of Medical Informatics in Geneva University Hospitals and School of Medicine, following in the steps of Prof. Jean-Raoul Scherrer who founded this world-renowned group.

His current research focuses on the development of innovative computer-based tools for improving the quality and efficiency of care processes, at the local level of the hospital, the regional level of a community healthcare informatics network, and at the global level with the development of a south-south telemedicine network in Western Africa.

Ousmane Ly is the Executive Coordinator of "Keneya Blown", the technical structure of Mali Network of Information and Medical Telecommunication (REIMICOM). Dr. Ly has a PhD in Medicine and a BSc in Biological Sciences. Currently, he is preparing his Post Medical Computing University Certificate in the University of Geneva. Mr Ly is also a member of the ATAC (African Technical Advisory Committee / United Nations Economic Commission for Africa, ECA).

REFERENCES

Geissbuhler A, Ly O, Lovis C, L'haire JF. *Telemedicine in Western Africa: lessons learned from a pilot project in Mali, perspectives and recommendations*. Proc AMIA Fall Symp. 2003:183-8

Graham LE, Zimmerman M, Vassallo DJ, et al. Telemedicine--the way ahead for medicine in the developing world. *Trop Doct 2003*;33:36-8

Oberholzer M, Christen H, Haroske G, et al. Modern telepathology: a distributed system with open standards. *Curr Probl Dermatol 2003*;32:102-14

Networked economy
Effects on organisational development and the role of education

Mikko J. Ruohonen
Department of Computer Sciences, University of Tampere, FIN-33014 UNIVERSITY OF TAMPERE, Finland

mr@cs.uta.fi

Abstract: The knowledge-based, networked economy demands new forms of collaboration between experts and organisations. Different domains of knowledge need to be integrated in order to create a new multidisciplinary approach for managing the challenge which we here call 'e-thinking'. Knowledge-based change is facilitated by the platform of modern ICT (Information and Communications Technology) and demands new and innovative concepts from the domains of e-business, e-work and e-learning. In support of a comprehensive e-thinking evolution at least these three domains of expertise need to be understood. Problems with the implementation of each of those "e-domain" approaches will be discussed and the rationale for integrating them is emphasised. Successful implementation of knowledge-based professional development is argued to be the alignment of organisational objectives, understanding of work and business processes, and knowledge about learning at work.

Key words: e-business, e-learning, e-work, organisational development

INTRODUCTION

Following the hype of the digital economy a large number of "dot-com" and also e-learning companies have suffered a serious downturn. Investors have not been satisfied with the financial results of this industry. At the same time problems with implementation of ICT-based processes in working life are being reported. Critical observers say that the development of our knowledge society has just been the development of technology platforms, without taking account of processes and contents.

One of the problems might be that comprehensive change of organisational activities relies both on organisational learning and appropriate use of ICT. Hence, a combination of technical and social understanding is needed. In the following we evaluate new phenomena in our knowledge society such as e-business, e-work and e-learning and discuss the integration of these domains.

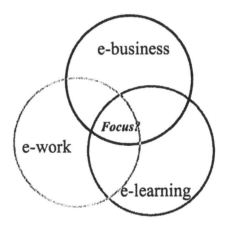

Figure 1. Focus of Knowledge Society development

REALISING INTER-ORGANISATIONAL COMPETITIVE ADVANTAGE

The use of ICT in organisations has emerged as an important managerial issue. In addition to improving cost-effectiveness, it also provides great potential for innovations and organisational change (Ruohonen & Salmela 1999). The e-business imperative is guiding the change process in which organisations must think about operational models, organisational processes and new ways to serve the customers. Functional hierarchies turn into organised networked structures. Advanced collaboration and outsourcing strategies are gaining importance (Kern & Willcocks 2002). However, collaboration is not just networking of transactions, it is also implies sharing of knowledge, uniting and integrating processes, and development of joint performance measures. These structures can be described as "knowledge management (KM) networks" (Ruohonen & Salmela 1999; Warkentin et al. 2001).

Products and services, involving ICT and digital contents, provide opportunities for new innovations, provided that a strong commitment and an ability to network reliably with partner organisations are present. The building of knowledge networks demands new qualities both of organisations and of their managers. Dyer and Singh (1998) have proposed four quality factors which are important for inter-organisational competitive advantage:

1. relation-specific assets, i.e. investment in the specific relationship of partners;
2. knowledge-sharing routines between partners;
3. complementary resources supplementing core competencies of each partner; and
4. effective governance of the relationships

Knowledge creating companies should manage all four qualities, especially so in situations where the company either cooperates within an ICT-based system, or has a constant outsourcing service relationship with a partner company.

Relation-specific assets

Co-operation relationships between organisations are not just formal agreements or just aimed at price control; organisations more and more look for long-term cooperation. For instance, outsourcing of services often is started on the basis of issues as pricing of services, cost control and exit provisions which provide a good start for closer cooperation with many options to change this relationship. The challenge is that the service provider organisation must be able to learn the business model and processes of the customer organisation. Customers on the other hand have to be ready to deal with products and service alternatives on a deeper level than transaction cost. The service organisation should take account of the customer's schedules and timetables. Commitment is needed to invest continuously in such a relationship, both economically and mentally. The success of the cooperation is especially strengthened by mutual trust and social exchange which is claimed to be a necessary condition to the genuine cooperation project (Kern & Willcocks 2002).

Knowledge-sharing routines between partners

Some new branches build their information system services behind web-based business service portals. In many cases these portals are just cyber-crossroads, i.e. the customer "drives" to the crossing, selects a new road, drives on and never returns. It is far better to have shared knowledge creation where the exchange of knowledge is a two-way process. Stakeholders must learn from each other to improve service for the customers and for better competitiveness. A dominating partner often uses such portals as a bulletin or bill board. However, from the problem-solving point of view a working portal requires a common knowledge sharing practice of the partners and customers. Many companies that have operated in this way, have a long time service development towards pro-activity, right pricing of services and anticipation of customer needs. The added value is not just control of transaction costs or reduction of intermediaries as was argued in the early days of e-commerce. There is a search for new value creation models.

Complementary resources or capabilities

Resource-based theory of businesses focuses on resources which are valuable, rare, difficult to copy and effectively organised from a competitive point of view. The core competence of a business is often created in the long run, self-steered and sometimes even by chance. Particularly for the management it is difficult to see the importance of competencies created on the business floor. Working knowledge is very often emerging from working and business processes (Davenport & Prusak 1998). However, these competencies can become obsolete, if competitive or technological changes affect our business environment. An inter-organisational setting which is based on learning in alliances (Larsson et al. 1998), also provides a good setting for developing competencies. Competitive advantage of knowledge networks is created through the clustering process in which two or more organisations with complementary competencies begin to compete against other competitors' clusters with similar interests (Ruohonen et al. 2003a).

Effective governance of relationships

Inter-organisational competitive advantage will not be achieved if the costs of cooperation and governance exceed the level of other competing clusters. Governance practice must be created, while different stakeholder groups control their own interests. Very often an independent mediating organisation or an official broker manages the service portal, based on agreed partner relationships. However, the growing role of context-related knowledge makes the contribution of an "independent broker" problematic. It is a dilemma both to stay neutral between collaborating partners and to become deeply knowledgeable on substantial industry issues. Practice implies that just delivery of contents is not enough, you also need to manage community building and develop and share contextual knowledge. The operational actions, appointed to particular persons in charge of the organisation, and the processes must be clear and explicit to every partner. The broker must be able to operate fast without constantly arranging meetings when the cooperation proceeds and the customer interface changes. Situations in which this customer setting has to be evaluated and checked will be created continuously.

E-WORK: COMMUNITY OF PRACTICE VIEW

Knowledge is created within communities

The problem in current developments of knowledge society is that knowledge creation is seen as technology-driven process and not as a work-driven process. Community of Practice (CP) experts emphasise contextual development of organisations (Wenger 1998, Brown & Duguid 2001). However, many organisations still rely on technology-driven knowledge creation and numerous databases (Wenger 1998). Contextual knowledge involves interactions, conversations, actions and interventions. Integration of working and learning demands focusing on formation and change of working communities. In organisations people acquire not only explicit knowledge, but also encultured knowledge to be able to act as a community member (Blackler 1995, Davenport & Prusak 1998, Brown & Duguid 2001).

A CP consists of three basic elements (Wenger web portal 2003):

a) *What domain it is about*—the domain of knowledge that gives members a sense of a joint enterprise and brings them together. The joint enterprise reflects the members' own understanding of their situation.

b) *How it functions as a community*—the relationships of mutual engagement that bind members together into a social entity. Members learn with one another and interact. The degree to which any group is a community of practice depends on how they function together; this cannot be decided in the abstract.

c) *What capability its practice has produced*—the shared repertoire of communal resources that members have developed over time through their mutual engagement. Communal resources include, for example, routines, lessons learned, sensibilities, artefacts, standards, tools, stories, vocabulary and styles. This repertoire embodies the community's accumulated knowledge.

CPs can easily cross formal organisational boundaries. Therefore a broader context should be considered of social learning systems, such as industries, regions, or alliances comprising multiple communities in interaction (Wenger 2003). Relationships among members are based on collegiality, and the community's purpose is to develop knowledge, not to allocate resources or manage people in order to deliver a product or service to the market. This multiple membership is crucial to the creation of a learning process that connects the development of knowledge and the work (Wenger 1998, Wenger & Snyder 2000). Managers should focus on bringing the right people together, identifying potential communities that will enhance the company's strategic capabilities, providing an infrastructure in which communities of practice can thrive, and using non-traditional methods to assess the value of the company's communities of practice (Wenger & Snyder 2000).

Miscommunication and misunderstanding are common at boundaries, because different communities have different ways of interacting. This creates a space for intercommunity learning in boundary processes and for the production of new knowledge (Wenger 1998). Multiple membership allows people to act as knowledge brokers across boundaries. Boundary objects can accommodate multiple perspectives, boundary activities, interactions and practices that force people of various communities to confront their experiences and perspectives. Technology platforms can make communication across boundaries easier. Boundary objects not only serve to coordinate, but to record and signal changes in one own community's practice (Brown & Duguid 2001).

Stewarding knowledge in communities of practice

CPs can vary in the extent to which they explicitly undertake the stewarding of knowledge (Wenger 2003). CPs might just be content to exchanges tips and lessons learned on an ad hoc basis. Some communities take responsibility for establishing and developing their practice and their community. Communities can evolve to be strategic in their thinking, explicitly viewing the development of their practice as a strategic move on behalf of the organisation. And in the last step some communities undertake to transform the organisation with the insights and new practices they have generated.

These levels do not represent a progression toward an ideal state. Each level has its value and is appropriate for some communities. But it is useful to see the range of what is possible and to be aware of the issues that communities face as they transition from one level to another. Understanding the value of CPs as stewards of knowledge is not always easy because the effects of community activities on performance are often indirect.

E-LEARNING IN THE CONTEXT OF WORKING ORGANISATIONS

Converging challenges of e-learning and e-training

In the IFIP TC3 E-training conference (Nicholson et al. 2004) it was stated that in the case of e-training and e-learning, many instructional programs are transmissive, viewing learning as passive, and focusing on individual learning, rather than as interactive and engaging — important attributes for the promotion of higher order learning.

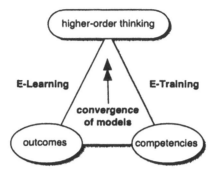

Figure 2. Convergence between E-learning and E-training models (Nicholson 2003)

Refocusing on the attributes of the underpinning learning models allows cognitive perspectives. A blended solution of cognitive strategies to support interactive and social learning is needed to maximise the potential for optimal learning and training. Nicholson (2004) states that we can expect to see a convergence of models as these both attempt to achieve similar kinds of outcomes (e.g., higher-order thinking) with an increasing focus of e–training on the knowledge-era workplace and of e-learning on real-world contexts. (Figure 2.) Nicholson suggests a new emphasis on developing effective learning models for both school and workplace learning, with the essential differentiation being in the context rather than the methods used to achieve learning outcomes, i.e., that cognitive aspects of learning should be central to both education and training models.

Learning in working life context

As noted many of the e-learning activities have been conducted in the school setting. However, a growing need is in the area of organisations and working life.

According to Järvinen & Poikela (2001) the most important task of management is to create an environment supportive of group activity, innovation and creation of new knowledge. However, there is also a growing discrepancy between cost-effectiveness pressures and mastering of innovations and knowledge creation.Finally you need to act as an acrobat to balance total quality and learning management (Ruohonen et al. 2003b).One of the opportunities is to foster informal and incidental learning at the workplace. Järvinen and Poikela (2001) propose a process model of learning at work (see Figure 3.) based on experiential and organisational learning models.

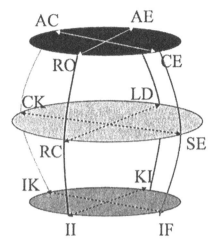

Figure 3. The process model of learning at work (Järvinen & Poikela 2001)

CE= Concrete Experience	SE= Sharing of Experience	IF= Intuition Formation
RO= Reflective Observation	RC= Reflecting Collectivity	II= Intuition Interpretation
AC= Abstract Conceptualization	CK= Combining New Knowledge	IK= Integration of Interpreted Knowledge
AE= Active Experimentation	LD= Learning by Doing	KI= Knowledge Institutionalization

Management should be aware of four different processes: social, reflective, cognitive and operational, which bind together individual, group and organisational learning. Experiential learning in a work organisation can be refined into a process description in which these processes follow, influence and shape each other in a process of continuous learning. Järvinen and Poikela (2001) state that this new description brings a new kind of modelling, in which an organisation is seen as being composed of processes instead of being seen as levels and hierarchies. Ruohonen et al. (2003) propose that many of the social and reflective processes are neglected in the current working life. The use of ICT enables and even dominates cognitive processes, such as intranet and other communication media and furthermore operational processes, such as enterprise and logistics systems. However, knowledge creation also demands social and reflective processes in order to foster social innovations.

CONCLUSION

It was stressed that a comprehensive understanding of different domains of "e-thinking" knowledge is needed to allow focused professional development of business, work and learning. Quite a lot of effort has been put into development of technology platforms. However, after the bursting of the hype-bubble, the nature of work communities and work-related learning are again highly relevant.

We have to avoid unrealistic top-down planning, while communities of practices are driving the knowledge creation process in which the members of the community decide on boundary objects between organisations or even alliance structures. In knowledge society most of learning is in the real-life setting i.e. incidental or informal learning. We need to understand more systematically the nature of learning at work. Otherwise we will loose sight of how development should go, either by focusing too much on top-down approaches or by just believing in liberation of work or by basing ourselves on technology-driven school pedagogy.

To be effective in engineering the Knowledge Society, our development efforts need to influence the three domains of knowledge, i.e. e-business, e-work and e-learning, Development slogans such as digital competence development or network learning (see Nicholson et al. 2004) confirm these ideas and show that some organisations are already ahead along these lines. Now we need to take it to the development agenda at society level.

BIOGRAPHY

Dr (Econ) Mikko J. Ruohonen is professor at the University of Tampere and a teacher at the Turku School of Economics and Business Administration in Finland. He has worked in the field of information resources strategy since 1984. Besides his teaching and research on information strategies, electronic business, knowledge management and inter-organisational learning he is active in several management executive programs such as the Executive MBA TIJO Program for CIOs and the Pori Management Development Program by the Tampere University of Technology. He has published four textbooks, over a 100 articles and lately prepared an e-business strategy report for the Finnish technology industry. He is the Chairman of International Federation of Information Processing (IFIP) Working Group 3.4. and Chairman of the Board of the Center for Extension Studies at the University of Tampere.

REFERENCES

Blackler, F. (1995) Knowledge, knowledge work and organisations: an overview and interpretation. *Organization Studies* 16(6), 1021-1046.

Brown, J.S. & Duguid, P. (2001) Knowledge and organization: A social-practice perspective. *Organization Science* 12(2), 198-213.

Davenport, T.H. & Prusak, L. (1998) Working knowledge: how organizations manage what they know. Harvard Business School Press, Boston.

Dyer, J.H. & Singh, H. (1998) The relational view: Cooperative strategy and sources of interorganisational competitive advantage. *Academy of Management Review* 23(4) 660-679.

Järvinen, A. & Poikela, E. (2001) Modelling reflective and contextual learning at work. *Journal of Workplace Learning* 13(7/8), 282-289.

Kern, T. & Willcocks, L. (2002) Exploring relationships in information technology outsourcing: the interaction approach. *European Journal of Information Systems* 11, 3-19.

Larsson, R., Bengtsson, K., Henriksson, K. & Sparks, J. (1998) The interorganisational learning dilemma: Collective knowledge development in strategic alliances. *Organisation Science* 9(3), 285-305.

Nicholson, P. (2004) E-Training or E-Learning – towards a synthesis for the knowledge-era workplace. In Nicholson, P., Thompson, B., Ruohonen, M. & Multisilta, J. (eds.) *E-training Practices for Professional Organisations*. Kluwer Academic Publishers.

Ruohonen, Mikko & Salmela, Hannu (1999) *Yrityksen tietohallinto*. Edita Oy, 1999. (218 pages, an information management textbook in Finnish)

Ruohonen, M., Riihimaa, J., & Mäkipää M. (2003a) "Knowledge based mass customization strategies – cases from metal and electronics industries" A paper presented at the "Mass Customization and Personalization Conference (MCPC) October, 2003, Munich, Germany.

Ruohonen, M., Kultanen, T., Lahtonen, M., Rytkönen, T. and Kasvio, A. (2003b) Identity and Diversity Management for New Human Resource Approaches in the ICT industry. In F. Avallone, H. Sinangil & A. Caetano (eds*)* *Identity and Diversity in Organisations*. Milan: Guerini Studio.

Warkentin, M., Sugumaran. V., & Bapna, R. (2001), E-knowledge Networks for Inter-Organizational Collaborative E-Business. *Logistics Information Management*, Vol. 14(1/2), 149-162.

Wenger, E. (1998). *Communities of practice. Learning, meaning and identity*. Cambridge University Press.

Wenger, E. (2003). Communities of practice. Stewarding knowledge. See www.ewenger.com checked 29 Nov 2003.

Wenger, E. & Snyder, W.S. (2000). Communities of practice: The organizational frontier. *Harvard Business Review*, Jan./Feb. 2000.

Understanding and interpreting the drivers of the Knowledge Economy

Mohan R. Gurubatham
International Business School, University Technology Malaysia, Level 2, IBS Building, Jalan Semarak, 54100 Kuala Lumpur, Malaysia.

mohang@streamyx.com

Abstract: The Knowledge Economy (K-economy) is much heralded as enabling the death of distance, the opportunities and promise of human capital development *via* life long learning and e- learning, the development of learning communities and knowledge enrichment of communities through community portals, to mention a few of the implications. It is certainly quite obvious that the K-economy is much more, than just technological software or hardware. The *enablement* of knowledge acquisition and utilisation, so that information can be effectively, efficiently and meaningfully transformed into wisdom, is examined along two fronts: 1.
Which drivers induce the diffusion and adoption of   ICT globally? What attitudes and competencies facilitate or impede the adoption process? The notion of *cognitive literacy* will be examined in this context.   2.
What are the needs and wants of knowledge societies that can be facilitated as *design features* for learning by understanding the subtleties of the dimensions of culture both from national and organisational perspectives?   An eco-textural paradigm is used to frame the discussion of the integral role of enabling technologies, in learning, personal and cultural enrichment. It is envisaged that the role of both affective and cognitive dimensions will be validated in *wisdom attainment* as the fulfilment of the Knowledge Society.

Key words: cognitive literacy, knowledge economy drivers, schema, transcend, value adds

KNOWLEDGE ECONOMY'S TECHNOLOGICAL DRIVERS

So what is exactly driving the knowledge economy? The answer first appears to be an interaction of the three technological drivers of the knowledge economy. These are specifically:

1. Moore's Law – the maximum power of a processing microchip at a given price doubles every 18 months. So What? Computers become faster while the price for a given level of computing halves.
2. Gilders Law - the total bandwidth of communication systems will triple every 12 months. So what? There is an approximately 40% decline in the unit cost of the net
3. Metcalf's Law - the value of a network is proportional to the square of the number of nodes. So what? With network growth, the value being of connected grows exponentially and costs per user holds constant and ultimately falls.

Induction and diffusion of the drivers

As computing power practically doubles, affordability becomes a more salient albeit promising dimension in providing access to the technology. Hardware adoption rates may seem staggered along the specific types of computing hardware enabled by the promise of Moore's Law until the end of this decade.

Business Drivers behind ICT adoption

Next, we outline the major business developments of the technological drivers of the knowledge economy discussed thus far. Behind all of these drivers is the explosion of information and communication enabled by the ubiquity of ICT.

Guiding Principle 1: The growth of international business where globalisation is inevitable.

– Perfect competition is increasingly real. Consumers are price savvy for any product anywhere by the 'death of distance'. There is demand to find new markets overseas. New markets are opened up anywhere and anytime. 24/7 customer service cycles such as call centres in India are now commonplace by business to consumer electronic commerce. With the advent of direct broadcast satellites and digital service lines, consumers are increasingly cognisant of consumption and lifestyles of their peers elsewhere. Expectations are engendered for price, quality and service.

- There is demand for quality management systems from overseas markets with higher customer expectations of value added. There is required agility to respond to evolving customer niches. For example, Southwest airlines business success has inspired similar low cost strategies in the EC by Ryan Air and even South East Asia by the emergence of Malaysian -based Air Asia.
- The deregulation of markets for *entree* by international firms into domestic markets have seen restructuring, reengineering and right sizing in the financial services and telecommunications industry. Other industries are not immune. There is increasing competition for presently protected monopolies. China after its WTO *entree* dramatically challenges other counties on cost and quality, or to climb the value chain ahead of it. Within China alone, expectations of consumers are being raised. China's Internet penetration rate is 5% or 56.6 million (A.C. Nielsen 2002). India while far lower than China in FDI is increasingly capable of competing in branded services and Information Technology Enabled Services (ITES). In terms of outsourcing these surging, not emerging economies represent competitive lower cost advantages.

Guiding Principle 2: There is an evolution from 'hard' value to 'soft' value where knowledge holds a disproportionate premium as an asset. Knowledge is essentially a mental product.

- The value of intellectual capital is greater than raw materials, manufacturing assembly and infrastructure alone. Nations have to climb up the value chain in of higher value added in technological innovation and creativity.
- Previously Neo Classical economics only recognized 2 factors of production: labour + capital.
- Now, New Growth Theory includes technology and has basis in Knowledge as intrinsic to the economic growth (Romer 1986; 1990). New Growth predicts technical platform effects and non-rivalry, that can lead to increasing returns on technology investment such as in 'open source' advocacy.

- New technology multiple potential effects for further platforms, is not one-off. Promise of diminishing returns is possible. Technology thus can raise Return on Investment. So developed economies can sustain growth.
- By traditional economic predictions there are diminishing returns. Developing economies even with unlimited labour cannot attain growth.
- Even in sectors such as manufacturing and government, ICT has ramifications. Just-In-Time-Information (JITI) informs the right person at the right time; from here knowledge workers must transform data into knowledge and application. Web-enabled business transactions drive speed, cost and partnering expectations. Scanning best practices, relearning and reinventing if necessary to stay ahead competitively must enliven the continuous and discontinuous process of shortening lead times. From performance management in well run airlines to global consulting firms as well 'old economy' manufacturing, knowledge management has caught on as in the case of Buckman Labs in the USA and Cemex in Mexico.

Guiding Principle 3: Soft knowledge can transcend 'hard' physical boundaries; ICT as process and not just infrastructure, enables the ubiquity of information.

- The death of distance is ubiquitous. In China, the proportion of internet users in a city is not determined by its size and stage of development. Chinese urban dwellers see it positively, as a virtual meeting place albeit in China only (Guo Liang 2004).
- The death of distance advantage is leveraged by the firm that has the greatest value-addition, the lowest 'weight' i.e., an intangible factor that transcends the 'weight' of industrial products such as from manufacturing. An example is software or solutions such as by INFOSYS, an Indian firm which is now a global brand name performing in markets such as the United States and Japan, and a quality performing stock in the NASDAQ, USA.
- Human capital investments are mandated to leverage the discontinuous improvements in Information Technology Enabled services (ITES). Relatively 'poor' countries such as India have surprised the world with their prowess in ITES. Enabled by the emergence of high bandwidth telecommunications networks, offshore business processing has indeed proved not just very viable, but also profitable. The range of Indian ITES services include the back-office processing of financial services, data mining and call-centres. Indeed a study by NASSCOM-McKinsey (2002)

has predicted at least 77–80 billion USD worth of revenue by 2008 in these services.
- ICT raises the profile of brands as perceptual assets. Brands are *schema* or mental short hands. Brands are critical because they reinforce consumers' trust. Brands transcend the functional attributes of their products but remain integrated within an *ecosphere* that reinforces consumer lifestyles and belief systems e.g., SKALI.net in Malaysia.

These drivers are related and fluidly interact to create a dynamic landscape of trends and events that have both intended and unintended consequences. Results of these drivers necessitate human capital investment to enhance the competitiveness of the workforce. ICT is thus the enabler. Next we look at the role of ICT in human capital development.

HUMAN CAPITAL DEVELOPMENT AND ICT

Firstly, factors that promote successful ICT adoption in organisations are:
- Strategically mandated organisational learning. Senior management must sponsor and relentlessly communicate that learning by trying out is alright. Emergent learning drives emergent strategies that better shape organisational resilience. Heuristics as opposed to algorithmic planning is more suitable for turbulent environments where organisations need to thrive or survive in uncertainty. Such is the environment that we currently face.
- Communication is part-and-parcel of the process of knowledge sharing. Knowledge management is doomed for failure, if recognition and reward systems are not in place for populating and accessing databases.
- The practical application of this value adding chain of information literacy in the management of the cultural environment demands not only empowerment, but managed responsibility. In the management of cognitive development, 'mindfulness' (Langer, 1997), and the sustenance of mindfulness (Senge, 1999), support mechanisms must be in place both formally and informally.
- Alignment of work roles to mission critical competencies so that career paths are defined against organisational goals. This signals that organisations are viable and desire longevity through adaptation beyond a short-lived project mindset.
- A transparent process view of value addition across work roles that diminishes silo thinking.

- Organisations that stress functional competencies still structure most departmental units as 'silos'. This is common in developing countries. To the extent that silos dominate, an overall and transparent value-chain 'process' mindset that transcends 'passing-the-buck' and finger-pointing of individual egos and departments is unfortunately dissipated or lost even with all the e-enablement. What are required are broad comprehension, sharp focus and commitment to teamwork imbued with a lively sustained sense of organisational mission and alignment.
- Active sponsorship, either morally or financially, for lifelong learning that enriches both work and play so that good performers bind to the organisation.
- Understanding that knowledge work involves the active manipulation of symbols to yield knowledge and ultimately wisdom. This is a process of human development that requires education, coaching and mentoring. With the adoption of ICT, the technology enhances this with accessibility, speed, reliability and scalability of data.

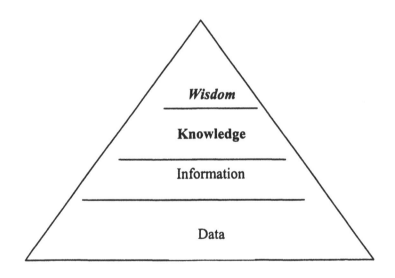

Figure 1. The Cognitive Literacy Chain

THE COGNITIVE LITERACY VALUE CHAIN

In the hierarchy of knowledge, data represent the lowest level of information value. The distinctive competency at this level is accuracy and speed. This level should be automated by information technology requiring little or no vigilance by human operators. In the 21st century, this level of the chain has little competitive advantage.

Next, the level of information represents more active information processing in interpreting data. Typical activities at this level for example include quickly recognizing critical quality parameters such as in statistical process control, inputting unique customer data in Customer Relationship Applications (CRM) while on-line, and recognizing key or salient customer information in call-centre tasks. These activities call for competencies in perceptual abilities.

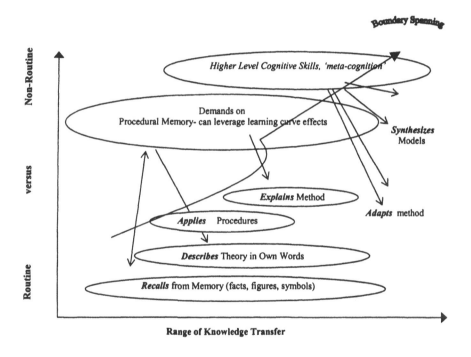

Figure 2. The range of knowledge transfer

Knowledge is higher-level cognitive activity, such as concept attainment, problem solving and innovation. Examples of such activity include coding (programming) software, applications, trouble-shooting, manpower and succession planning. Most of the routine administrative and management competencies that involve discrimination and judgment enabled by computer applications, such as Human Resource Information Systems (HRIS) or Enterprise Resource Planning systems (ERP), may be categorized at this level. Other examples of value-added thinking include the ability to extrapolate and interpolate data in market trends, even in the face of 'missing data or information' – breakthrough thinking is often experienced as this.

The uppermost level of the cognitive literacy is wisdom which is deeper and wider. Wisdom is essentially an evaluative and integrative mode of thinking that can subsume the lower cognitive levels of thinking. The ability to integrate and evaluate requires use of the 'lower' levels of mental activity such as perception and cognition, but more critically, its success rests on the ability to yield insight that derives from more than the sum-of-parts of data. This process, which has been described in hierarchies such as Bloom's Taxonomy, involves affective and ethical dimensions of human judgment. Examples include the ability to integrate and evaluate simulations and perform strategic scenario planning such as in market research involving data mining, multidimensional scaling, organisational modelling and so on. Often this value is driven by well-developed social cognition schema and by what is commonly dubbed as intuition. Ethical judgments predicated upon higher levels of moral development can safeguard against the abuse of the access of information. Cultural values and assumptions can subtly and powerfully influence evaluations.   Wider and deeper cognisance must encompass varieties of stakeholders.

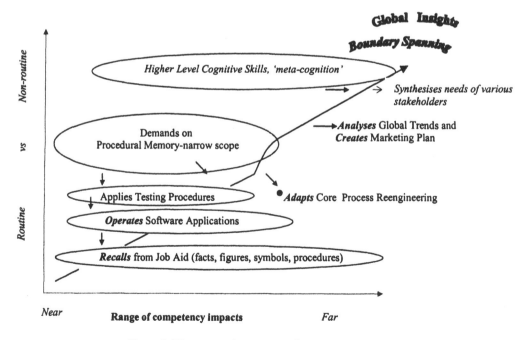

Figure 3. The range of competency impacts

If knowledge transcends physical boundaries, deeper knowledge transcends domains of function but capture overall process. Such knowledge is essentially higher value added or Key Value Added (KVA) i.e., how long does an average person require to learn a process as embedded knowledge. (Kannan &Akhilesh, 2002).

Learning that nudges higher value added thinking is dubbed 'far transfer' (Salomon and Perkins 1989). Such learning is insightful, and often based upon deep principles that transcend context or domain (Figure 2.). Its range of competency impacts transcends organisational boundaries (Figure 3.). Learning strategies that induce these include meta-cognitive, i.e. 'high road' strategies such as reflection, analogising by cases and boundary pushing, Socratic prompting, introspecting assumptions and values. Techniques that expand the conscious thinking mind such as Transcendental Meditation ™, indicate increased operationally defined wisdom and enliven the whole brain. Well-controlled and longitudinal studies have found that TM practice enhances a wide range of functions spanning the entire cognitive literacy chain, including developing shorter response latencies in discrimination tasks, increased field independence, increased flexibility of perception and improved verbal problem solving, increased and creativity, and increased fluid and culture fair intelligence. For a thorough review of published studies, see Alexander C. N. (1993).

Next, we look at the role of culture in cognition and ICT.

LEARNING, ICT AND CULTURE

Schemas are mental structures that guide human information processing. They are the basis of norms, parameters and expectations. Schemas are automated in long term human memory over time as 'scaffolds' that drive the top-down assimilation of information and file data into categories. Culture is schematised over long periods of time (Gurubatham 2001). Cultural schemas act as 'notional guidance systems'-'scripts' that guide behaviour in cultural settings, and can stubbornly and subtly influence higher-level thinking.

Cultural scripts may represent obstacles that impede or facilitate learning and successful adaptation to globalisation. Cultural adaptation typically does not commensurate with economic growth. Unbundling automated cultural schema is inherently stressful (Gurubatham 2001) What is needed is a way to preserve the ecological diversity of culture, harness it and transcend it for cognition of deeper, universally shared values and laws. An analysis of the major cultural dimensions will help to structure ICT interventions effectively.

The Dimensions of Hofstede's Ecological Factor Analysis (1980; 1991):

- Power Distance (PD) indicates perceptions of distribution of power in society. In organisations, High PD indicates hierarchical relationships, status effects, top-down communication and authority-compliant behaviour. Norms, rules and work cultures are top-down, and organisational members are socialised into this. Examples are Asian, African and Middle Eastern and 'Latin' societies.
- Low PD indicates flat organisational structures equality status, informality and horizontal flow of communications. Examples are largely western nations and organisations.
- Collectivism indicates the degree of group orientation from individualistic to high collectivist. Westerners by and large are individualistic compared to Asian, African and Middle Eastern and 'Latin' societies. Note a related concept of High and Low Context by Hall (1976). High Context (HC) is characterised by relationships, importance of non-verbal communications, warmth, and *polychronic* in time orientation i.e., multi-track and non-linear time perception. Low context (western-industrialised) is more digital' and emphasises verbal communication , rigid task orientation and punctuality. High HC may also be characterised by a natural alignment of the transcendent *Self* with the *ecosystem* and the *cosmos* (Gurubaham 2001). Intuition here is *refined affect* and may underpin wider and deeper judgement i.e., wisdom. Indeed, affect may be more archetypal and bypass typically 'western' cognitive activity (Zajonc 1980).
- Masculinity versus femininity indicates a continuum of separating gender role definitions. Masculine cultures prefer rigid gender roles and are stress-confrontative and stress-achieving. Feminine cultures stress nurturance. Japan, Austria, the USA and Arab countries score high in masculinity, whereas the Scandinavian counties score low.
- Uncertainty Avoidance (UA) is the degree of risk aversion where high UA avoids risk and low accepts and even welcomes it. High UA cultures tend to be expressive, desire structure, simplicity and formality, are more likely to resist change and tend to be more proactive than ad hoc. Most Asian countries are high UA except Hong Kong which scores the lowest globally. Low UA favours informality and relaxed business arrangements, lives day-to-day with loyalty not being paramount.
- Long Time Orientation is simply a culture's reverence for patience, social trust and unequal social relations. Originally a Confucian concept, but is somewhat confounded in globalisation as Chinese societies are rapidly modernising today. Even in smaller Chinese cities,

there is a major perception among internet users who see it as a tool for freer political discussion and view its content with social trust (Guo Liang 2003), counterbalanced by a wider social need to control it.

CONCLUSION AND *ECOTEXTURE*

Since culture evolves slowly and is stubborn a wise guiding principle is to work with it rather than change it. Accept it as part of the *ecotexture*.

ICT Design Features by Non Western Cultural-Considerations

– Higher Power Distance (PD) will naturally prefer a relatively higher degree of information structure than lower PD. Control and access issues are very important with centralised power structures. Mental models or *schema* may favour hierarchical relations.

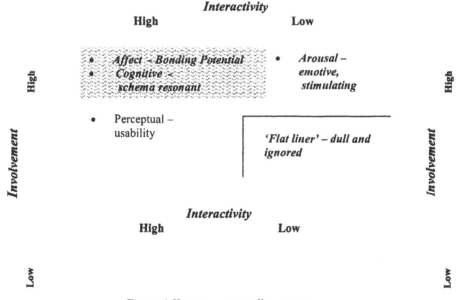

Figure 4. K-economy e-media process

Further, high PD may inhibit participative management that solicits inputs from workers by their formal hierarchies. Also favours official insignia,

logos, and content certification. Formal status effects are active that may inhibit for example a professor's acceptance in chat rooms.
– In High Context (HC) collectivist e-learning and knowledge management, courtesy, personal acknowledgment, ritual and relationships can be infused into the content to counterbalance the formality of power distance. High context cultures are naturally coherent and are informal in communications. ICT should complement this trait. Affect in HC is enlivened this way. Some of the elements of polychronicity may be exploited and synergised over time to accommodate a higher degree of task orientation for multitasking such as in call centres, utilizing on-line performance support while attending to customers.
– Masculinity and femininity – feminine cultures may be more comfortable with content ambiguity in gender roles. Masculine content may be biased towards task, mastery, sport, and utilitarian features.
– Long Time Orientation content – content, interactivity and involvement must focus on personal relationships as trusted information sources, patience in achieving results - usually social goals, and pragmatic value (Guo Liang 2003).

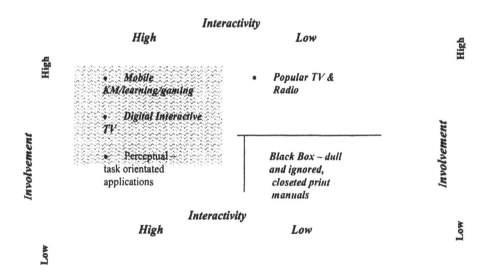

Figure 5. K-economy media genre

– High Uncertainty Avoidance (UA) may also inhibit 'discovery' (inductive) learning and needs prompts and debriefs until confidence is acquired. 'Advance organisers' (concept maps), visual aids or metaphorical, culturally endeared phrases may be used to provide HC and UA learners with navigation for complex content. Learner control may be initially inhibited and requires active coaching in the initial stages of e-learning. High UA may also inhibit populating databases in knowledge management. Counterbalance by enlivening the high context of teamwork, sharing and interpreting information. Level of choice and amount of data should be low. Keep out superfluous links. A help system is better if it is task driven rather than content driven.

The far transfer of knowledge establishes a guiding principle that ICT must strive for. Design features must include interactivity (usability) and especially deeper involvement (psychological bonding inherent in the culture) as seen in examples in Figures 4 and 5.

The use of multiple scenario simulation outcomes in content that is culture-varied and rich, with inductive principles complemented by coaching, can nudge the 'far transfer' of key principles that transcends context, yet appreciates diversity and engenders wisdom.

New 'wisdom technologies' such as Transcendental Meditation programs are "Natural Law' based. TM research indicates attainment of psychological stability even in the face of stressful environmental stimuli (Orme-Johnson 1973) so that individuals and groups enjoy both progress and their cultural heritage. In this technique, individuals practically experience their Self as the Unified Field underlying all cultural diversity and change (Chandler et al. 2004; Broome et al. 2004).

BIOGRAPHY

Mohan Raj Gurubatham is interested in the cultural and cognitive bases of learning and strategy in advanced Asian countries especially with pressures of globalisation in the knowledge economy. He is currently integrating the role of intangible factors in ICT within organic 'eco-field' frameworks that will shape practical consequences.

REFERENCES

Alexander C N (1993). *Transcendental meditation*. In Corsini, R J (Ed.), Encyclopedia of Psychology, (2nd ed.), New York, Wiley Interscience.

Broome, R., Orme-Johnson, D. W., & Schmidt-Wilk. J. (in press) Worksite stress reduction through the Transcendental Meditation program. *Journal of Social Behavior and Personality.*

Chandler, H. M., Alexander, C. N., & Heaton, D. P. (in press). Transcendental Meditation and postconventional self-development: A 10-year longitudinal study. *Journal of Social Behavior and Personality.*

Guo Liang (2003) The Cass Internet Report 2003: *Surveying Internet Usage and Impact in Twelve Chinese Cities,* Chinese Academy of Sciences, Beijing.

Gurubatham, M. R. (2001) *Maximising Human Intelligence Deployment in East Asia-The 6th Generation Project,* Palgrave, London.

Hofstede G. (1980) *Culture's Consequences: International Differences in Work-related values.* Beverly Hills: Sage

Hofstede, G. (1991). *Cultures and Organisations: software of the mind.* London: McGraw-Hill.

Hall ET (1976) *Beyond Culture.* New York, Anchor Books.

Kannan, G. and Akhilesh K.B. (2002). Human capital knowledge value added: A case study in infotech. *Journal of Intellectual Capital,* Bradford. Vol.3, Issue 2: pp. 167-180.

Langer Ellen J. (1997) *The Power of Mindful Learning.* New York : Addison-Wesley Publishing Company Inc.

Orme-Johnson, D.W. (1973). Autonomic stability and Transcendental Meditation. *Psychosomatic Medicine,* 35, pp. 341–349.

Romer, Paul M., (1986) Increasing Returns and Long-Run Growth. *Journal of Political Economy* 94(5), pp.1002-37.

Romer, Paul M., (1990) Endogenous Technological Change. *Journal of Political Economy* 98(5), pp. 71-102.

Salomon,G., & D. P. Perkins (1989) *Educational Psychologist,* 24(2), pp. 113-142.

Schmidt-Wilk, J. (2000). Consciousness-based management development: Case studies of international top management teams. *Journal of Transnational Management Development,* 5 (3), pp. 61–85.

Senge P, George R & Bryan S (1999) *The Dance of Change; The Challenges of Sustaining Momentum in Learning Organizations.* New York: Doubleday.

Zajonc, R (1980) Thinking and Feeling: Preferences Need No Inferences. *American Psychologist* 35 pp. 151- 175.

Beyond technology
Mankind as an end, or the end of mankind?

André-Yves Portnoff
Director of Observatoire de la Révolution de l'Intelligence at Futuribles, Tour Défense 2000, A296S, 23, rue Louis Pouey, F-92800 Puteaux, France, tel. 33 (1) 47730243.

andre-yves.portnoff@wanadoo.fr

Abstract: What we call a knowledge-based society is in reality a society based on intangible resources. Information is only valuable if it is structured in the form of knowledge that subsequently will impact society, when people decide to exercise their competences. Our society is dependent on all kinds of factors that determine our decisions and the efficiency of our actions. Passions and willpower are as much involved as knowledge itself. Creativity will become the critical resource restoring Man in a central position in society. Creativity cannot be mobilised by means of force. Therefore development is dependent on whether fundamental freedoms are respected. The digital revolution speeds up the evolution of the knowledge-based society. It may result in a form of development that might be aggressive to mankind and its environment. But it is also possible to promote a positive form of development, if we are able to show that the rules of the game have changed.

Key words: democracy, management, technological impact, society

WHAT IS IT WE ARE TALKING ABOUT?

When trying to build up this knowledge-based society, we have to ask ourselves what are we really talking about: what exactly is this new society that we associate with knowledge, information and communication, and even with intangible factors? The World Summit on the Information Society in Geneva has been labelled by governments as being on the "information society". The reason for this is that political leaders are impressed with the power of information technologies and by the transformations that are induced into our economy and our daily lives. However, this technician inspired vision implies that information has acquired an extraordinary value. In fact information has always been important, especially when rare and hard to obtain. But today information is more and more abundant and accessible. Its value is dropping, and even turning into the negative when spam or useless e-mail copies are piling up in our mailboxes. Information of value is the kind of information that has been sorted, interpreted, combined with other information in order to become knowledge.

Because of our competences, we are able to let various selected informations interact with one another which grants information sense and value. To build up knowledge out of information we must already have an appropriate kernel of knowledge at our disposal and a scale of values to help us sort data fulfilling our aims. And then we also need the will to complete the necessary efforts for the task. Some writers claim that, thanks to digital technologies, the transmission of knowledge is performed at no cost, since we can instantly duplicate, free of charge, any piece of work and send as many copies as we want to the other end of the world. This might be true when just sending digital files, but it is not true for knowledge which only becomes real when it is endorsed by humans. Human effort is needed to read the digital files, to understand it, digest it, and eventually possess its contents by transforming it. All these steps require time and involve many intangible factors, from intellectual faculties to desires and intuitions which are part of one's own personal culture.

Knowledge will only impact society if it is used and processed. There is a world of difference between data acquisition and competence. The latter is a capability to put knowledge to work. Competence involves character, behaviour, mind patterns and a way of thinking. For example, a medical doctor may have very deep knowledge of a speciality. But, he may nevertheless be a very poor physician, incapable of effectively treating his patients, because he has no integral view on their condition and is only focussing on his own speciality. This narrow and Cartesian way of thinking separates problems from one another and may lead to wrong diagnosis and prescription, ordering special treatment, efficient for arthritis but with dangerous side effects for the digestive system. To make things worse this medical doctor may have no social skills and will refuse to discuss the case with colleagues from other specialities, failing to gain necessary help. This professional will, more or less abruptly, produce a clever diagnosis triggering a psychosomatic reaction aggravating the whole situation.

Besides raw knowledge, real competence involves two extra elements: one intellectual and the other behavioural: a new way of thinking that is suited to deal with complexity, and a willingness to interact with others (Portnoff & Lamblin 2003).

Table 1. Three steps of accelerated mutations

	12th Century	18th & 19th Century	1950- today
	Medieval revolution	Industrial revolution	Intelligence revolution
Materials	Iron	Steel & concrete	Hyperchoice, nano-materials
Energy	Water-mills	Steam engines. Combustion motor	Harnessing energy
Life	Selection of plants and animals	Microbiology	Genetic engineering
Time	Hours (clock)	Seconds (chronometer)	Instant communication
Work organisation	Serfdom	Taylorism	Participatory and quality management
Nature of work	Repetitive and physical works	Repetitive and physical works	Creative and relational
Human status	Interchangeable labour	Interchangeable labour	Possessor of expertise
Society	Reproduction Society Slow changes except particular periods	Reproduction Society Slow changes except particular periods	Creation Society, constant changes
Arena of power	Land	Finance	Intangible: Mobilisation of talent, trust
Tools of power	Constraint	Constraint	Intangible : Conviction, emotion
Critical resource	Material resources	Material and financial resources, asymmetry of information	Intangible: shared information and power, creativity

NO SECURITY WITHOUT FREEDOM

The concept of "information society" puts too much stress on the information and computer side of it all. The real revolution is in the dramatic increase of available knowledge (Gaudin & Potnoff 1983; 1985). But it is rather unfair to talk about a knowledge-based society as if organisations till now have been based on ignorance! What is new? Traditional resources like raw materials or physical force are still important, but not critical any longer. Provided we can put human creativity to work, we will not suffer any shortage of natural resources or energy. If one of these happens to be missing, we shall have all the necessary knowledge to find a replacement. But we must take into account two new elements that might limit our efficiency. The first is that we have the power to destroy ourselves, if we lack wisdom and understanding. Our stocks of nuclear weapons are large enough to suppress every single form of life on earth. If we prefer a slower process of self-destruction, we must continue to seek short term financial profits, while neglecting the environment and international law.

We can also deny our citizens sources of information and expression, and prevent specialists from warning us of all the upcoming dangers. This happened in the former Soviet Union and the result of it was the Tchernobyl explosion. (Portnoff & Portnoff 1994; p. 38). This is still the case in today's China and the result is the out-of-control outburst of SARS and AIDS (Portnoff 2003). Nowadays, the exercise of freedom in an environment that respects the law, is the major condition to guarantee security.

NOT MORE EFFICIENCY WITHOUT RESPECTING OTHERS

The second new element is that in a fast and permanently changing world, innovation is crucial. This implies that all the time we recreate the very conditions that allow us to live in a modified environment. That is why we have suggested to use the expression "Creation-based Society" (Gaudin & Portnoff 1983; 1985).

Creativity has become the critical resource and condition of our development. And this makes freedom of thought, freedom of communication and freedom of expression so vital since creativity vanishes when one tries to repress these. Creativity has always fed on interaction between different cultures, on tolerance and on individual and collective freedom.

Innovation is more and more the outcome of a networked kind of collaboration. The growing complexity of all the issues and the rising number of competences that one has to put together in order to deal with these issues, obliges us to build collective forms of intelligence. This means that not one single person has all the necessary competences. Whether we are an individual team member, a corporation, a region or a country, in all cases we need to convince those who have complementary talents to join forces with us and collaborate. The goal is not so much to arithmetically sum up competences, but to build up synergy. The end result is a force that might be either far superior, or far inferior to the sum of individual resources that have been shared in the process, depending on the quality of the various interactions between players. Positive interactions are only possible between free and willing partners. It is possible to force other people to share with us their physical strength, but never their creative intelligence. The conditions to gain efficiency have switched from a logic of coercion to the logic of persuasion and seduction. We now have to convince others and win then into partnerships.

The efficiency of partnerships between corporations and of strong customer relationships has been demonstrated by the success of companies such as Wal-Mart and Dell. In his paper Mikko Ruohonen (Ruhonen 2004) shows the competitive advantage created by clustering two or more organisations with complementary competencies. Should this be extended to personal relationship? If everyone agrees that people are involved, the question is: which people? Should involvement be limited to high ranking executives, those "symbol manipulators" to quote Robert Reich, excluding all others and considering them has second-class material? We are convinced that the option of excluding the majority of staff does not provide the necessary reactivity in a fast moving environment and in front of a complex set of issues. We can no longer afford such a waste of intelligence and creativity, as we did when humans and machines where run by Taylorism.

The more hierarchical and divided an organisation, the less easy and fast communication and exchanges will be. This type of organisation will be the less innovative and reactive. We must never forget that the destruction of the space shuttle Challenger on January 28, 1986, was largely due to lack of respect in the hierarchy (Mayer 2003). The vice-presidents of the company who produced Challenger's booster rockets, refused to listen to experts that kept opposing the launch. They were mere employees and in such a hierarchically structured company, one was not used to listen to lower ranking staff members. This kind of behaviour is still rather common today and is one of the main causes of malfunctions within corporations and administrations.

THE NEW POWER OF CITIZENS

Anyhow, the situation is that nowadays, in developed countries, people are no longer willing to be considered as cattle or robots. Surveys in 1981, 1990 and1999 on values in European countries (European Values Study, EVS) (Arval & Futuribles 2002) have noticed a rise of individualism, a strong desire for self-expression and at the same time willingness to develop positive relationships with others. These trends are connected with strong ethical expectations. This is one of the reasons why no country in the world has ever reached the level of economical and technological development that we enjoy in the West, without undergoing democratic changes due to pressures from the masses. This has been the case in Taiwan and South Korea, following the footsteps of Japan. The digital network revolution amplifies this trend. The Internet acts as an accelerator in the process of spreading ideas, as an enabler in the search for information and in establishing collaborations. The cost and communication problems involved with activities, such as remote collaborative work and exchange of information, are dropping fast. Better informed citizens may exercise a choice, when confronted with commercial and political proposals. Whenever they are dissatisfied, they can express themselves over the Internet and set up efficient online consumer lobbies. We are witnessing the rise of a new power balancing traditional commercial, financial, ideological and political powers (Dalloz & Portnoff 2001). Never in history have small corporations and small territories had so many possibilities to express and defend their own personalities, provided they agree to get involved in partnership networking. Many non-governmental organisations use the Internet in order to protect the environment, minority rights, and more generally human rights. Next to Amnesty International, many lesser known organisations are fighting, especially in Latin America, to promote integration of women, rehabilitation of sexually abused children, freedom of press and education, and anticorruption laws. In her project, Annie Corsini-Karagouni (Corsini 2004) gives a convincing example of e-solidarity against female genital mutilations in Kenya.

Today, authoritarian regimes are disturbed by the kind of testimonies carried over the Internet. Obviously these new media, which are at the disposal of the people, can be used by all sorts of powerful organisations, including fast growing crime related ones. Many are seeking the role of Big Brother by using modern capabilities to control our lives, spread false information or to try intoxicate us with their commercial, political or religious propaganda. But, for as long as the world has existed, every single new means of communication has been simultaneously used by both promoters of individual freedom and by its fiercest opponents. Fortunately, since the printing machine was invented, every improvement in the speed of communication has resulted in promoting freedom, the last achievement being the fall of the Berlin Wall.

THE FUTURE IS WIDE OPEN

Instead of being frightened by the future because many dark scenarios are possible, we must be pro-active and work in favour of a humanist Renaissance scenario. This is certainly possible, if we really want it. First of all we must have the courage to say 'no'. There is no unavoidable fate, the future is not written anywhere and nothing binds us to accept that wrong political and economical forms of government may continue for the only reason that they have always been around.

For otherwise, this will surely lead us to economical, social, political and ecological major catastrophes as Amartya Sen (Sen 1999) did show and prove. We must try to input some form of intelligence into individual and collective egoisms and convince leaders of this reality: any organisation turns out to be economically more efficient, when respecting human aspirations and distributing responsibilities and power. This kind of organisation, whether it is a company or a country, will show a higher degree of creativity, will react faster to changes, and will know better how to handle the unexpected. Organisations that have retained Taylorism will be slowly outnumbered by those who set their money on networks and human beings.

We must clearly show to those who would be tempted to use the power of the Internet to control us, sell us any kind of goods and offend our privacy, that there is a red line that should not be crossed or otherwise they will provoke violent reactions from consumers and citizens. These misbehaviours would ruin confidence in electronic commerce and customers would soon disconnect from the network. It is most important that all the players understand the new rules of the game: we are all interrelated and interdependent.

Because of this we need a global way of thinking that helps us to understand the complexity of the issues. This complexity-based thinking must be developed and trained starting at school and this requires a higher level of cooperation between teachers to converge the various teachings and topics.

We must support « positive » technological options: peer to peer exchanges against hierarchical organisations, shared standards (open source) against proprietary solutions (Windows)

Finally, we can make the best use of digital networks within organisations to demonstrate how counterproductive traditional partitioning can be, isolating people and preventing them from seeking global efficiency. Water proof compartments do not always save ships, but they will surely sink us, if we do not remove them.

We must repeatedly demonstrate that in the long term sustainable development is the only solution for life. That freedom and democracy are necessary for creativity and safe technological progress. Knowledge without freedom is useless and dangerous: there is no example in the world of real technological and economical development at European level without education and democracy: Japan, South-Korea and Taiwan became democracies before reaching their present technological excellence. The two key words that both support human and economical development, and reduce the digital divide are "Human Rights"; two words, forgotten by Unesco in the WSIS draft final document. These include also Women's Rights, as Amartya Sen explains; the right for girls to go to school and for women to work outside the home.

BIOGRAPHY

André-Yves Portnoff is doctor in metallurgic sciences and director of the "Observatoire de la Révolution de l'Intelligence at Futuribles International". He is co-author of "La Révolution de l'Intelligence (1983-1985)", the first report that introduced the concept of the intangible based society in France. Journalist and consultant on foresight, he currently collaborates with large businesses and with SMEs interested in integrating the consequences of human and technological evolution into their strategy and management. His publications include:

- *La Révolution de l'Intelligence*, Science & Technologie1983-1985, The Intelligence Revolution, Gamma Institute Press, 1988.;
- Sociétés bureaucratiques contre Révolution de l'Intelligence, avec Arlette Portnoff, L'Harmattan, 1994 ;
- *Consommer, produire et distribuer en 2010*, avec Xavier Dalloz et Olivier Géradon de Vera (Gencod 2000) téléchargeable à www.dalloz.com .
- *Sentiers d'innovation, Pathways to Innovation*, (bilingual French and English edition), Perspectives-*Futuribles*, 2003.

ACKNOWLEDGEMENT

English translation: Jean-François Susbielle.

REFERENCES

Arval (Association pour la recherche sur les systèmes de valeur) (2002) *Futuribles* n° 277 numéro spécial, juillet-août 2002.

Corsini-Karagouni, Annie (2004) *E-solidarity, a means of fighting against FGM (Female Genital Mutilation)*. In: Weert, Tom van (2004) Education and the knowledge society, Information Technology supporting human development. Kluwer Academic Publishers, Norwell.

Dalloz Xavier et Portnoff André-Yves (2001) *L'e-novation des entreprises, Futuribles* n° 288, juillet-août 2001, pp. 41- 60.

Gaudin Thierry & Portnoff André-Yves (1983 & 1985) *La Révolution de l'Intelligence, Sciences & Techniques* ed., 1983 & 1985, Paris ; The Intelligence Revolution, Gamma Institute Press, Montreal ; La Revolution de la Inteligencia, INTI, 1988 Buenos Aires.

Mayer Paul (2003) Challenger, *les ratages de la décision*, PUF,2003 Paris.

Portnoff André-Yves (2003) *La censure, arme de destruction massive, Futuribles* n° 288, juillet-août 2003, pp. 67-70.

Portnoff André-Yves & Portnoff Arlette et al. (1994) *Sociétés bureaucratiques contre Révolution de l'Intelligence*, L'Harmattan, 1994 Paris.

Portnoff André-Yves & Lamblin Véronique (2003) *La valeur réelle des organisations, la méthode VIP, Futuribles* n° 288, juillet-août 2003, pp. 43-62.

Reich Robert (1991) The Work of Nations, Alfred A. Knopf, Inc. New-York.

Ruhonen, Mikko (2004) Networked Economy – effects on organisational development and the role of education. In: Weert, Tom van (2004) Education and the knowledge society, Information Technology supporting human development. Kluwer Academic Publishers, Norwell.

Sen, Amartya (1999) *Development as Freedom*, Alfred A. Knopf, Inc, New-York.

Social engineering of the Internet in developing areas

Wesley Shrum
*Louisiana State University, Department of Sociology, Baton Rouge, Louisiana 70803 USA,
Tel. 225-578-5311, Fax. 225-578-5102*

shrum@lsu.edu; http://worldsci.net

Abstract: Communication among researchers is fundamental to the development of
              knowledge in both developed and developing areas. Internet connectivity is
              now a precondition for participation in research communication. Establishing
              reliable and efficient connectivity at reasonable bandwidth is a task that is
              assumed to be relatively easy and straightforward in developed countries, but
              is surprisingly difficult in developing areas. Our project has sought to establish
              connectivity for university departments and government research institutes in
              India, Ghana, and Kenya but has yet to experience an unqualified 'success' for
              a variety of institutional and relational reasons. The concept of 'reagency' is
              used in preference to 'development' to explain the priority of personal relations
              introducing significant constraints that must be faced directly to establish
              connectivity in developing areas.

Key words: Africa, developing areas, Internet, science, technology

ENGINEERING THE KNOWLEDGE SOCIETY

"Engineering the Knowledge Society" is a large and beautiful concept that resonates with—indeed, underlies—the World Summit on the Information Society. My topic today centres on the Internet in developing areas, particularly in the educational and research sectors, and the ways in which we need to be particularly attentive to the character of the connectivity initiatives that are undertaken under the rubric of the "digital divide". If we are not careful, they will fall prey to the same problems of face-to-face interaction between donors and recipients that have characterised prior initiatives. What I want to emphasise before discussing our Louisiana Internet Project is the close correspondence between the idea of "engineering the knowledge society" and the approach generally taken by those in the field of "STS". (The Louisiana Internet Project is sometimes referred to as the World Science Project but that is a misnomer—it is because the domain name (http://worldsci.net) is a placeholder web page for a follow up event now being planned for the Tunis phase of the Summit.)

SCIENCE AND TECHNOLOGY STUDIES

"STS" once meant "Science, Technology, and Society" and now means, for many of us in this field "Science and Technology Studies". This is an interdisciplinary field with its own journals (*Science, Technology, and Human Values; Social Studies of Science*), professional society (the Society for Social Studies of Science), annual meetings, handbooks, and all the other apparatus one expects from a scholarly area. The reason that I bring this up is that this change of name—one should properly say this ambiguity in the meaning of the initials "STS"—reflects a long term shift in the way that scholars have come to view technology: not as a separate entity but as inextricably interwoven in the fabric of social organisation. Ideas that Technology Affects Society, or that Society Shapes Technology have become less significant than the notion of the "seamless web" in which things and people and social formations are tied together in networks or dynamic assemblages. That shift represents a growing awareness that technology is socially constructed, an awareness that many would equate with social engineering. The guiding idea here at our Geneva meeting is that ICT cannot be seen as a separate entity—and that is the close correspondence with the core theoretical and empirical concerns of STS.

RESEARCH COLLABORATIONS

Now may I qualify this in an important way—when we go on to say that the application of ICT has to be engineered, this must be viewed as the kind of interactive social engineering that is represented by research collaborations involving multiple countries, multiple organisations, and multiple points of reference. It would not be correct to view this kind of engineering as a direct-from-headquarters mandate to create a new product or design a new structure. The Internet is a collaboration in two respects:

1. As an innovative and powerful set of technologies, computer-mediated communication changes the conditions of physical presence and time delay that have characterised most forms of human interaction;
2. As an initiative focusing on connectivity, the Internet is the greatest collaboration the world has ever known, "greatest" both in the sense of potential benefits and potential danger.

That structure is often called decentralised but we must be careful when we apply that term to global diffusions such as the connectivity initiative that is the most obvious reference of "engineering the knowledge society". The connectivity initiative is no more decentralised than initiatives on poverty, education, agriculture, health, and environmental sustainability.

DIGITAL DIVIDE IN EDUCATIONAL AND RESEARCH SECTORS

After about ten years of work in the developing world, it became obvious to me that the digital divide is nowhere more evident than in the educational and research sectors themselves. That is ironic, of course, since it was within those organisations that most of the primary innovations occurred in the developed world, innovations that together became the Internet. Yet communication among researchers is fundamental to the development of knowledge in both developed and developing areas. Internet connectivity is now a precondition for participation in many forms of social involvement—like applying to my university for graduate work. Among these is research collaboration. Establishing reliable and efficient connectivity at reasonable bandwidth is a task that is assumed to be relatively easy and straightforward in developed countries, but is surprisingly difficult in developing areas. Our project has sought to establish connectivity for university departments and government research institutes in India, Ghana, and Kenya but has yet to experience an unqualified "success" for a variety of institutional and relational reasons. What we have discovered is the conceptual poverty of the notion of "development" as an adequate description for the processes we have witnessed since beginning our own connectivity project. We now use the concept of "reagency" to refer to the priority of personal relations introducing significant constraints that must be faced directly to establish connectivity in developing areas.

For the sake of perfect honesty, I must tell you that I am a sociologist and my primary teaching interest has been contemporary theory for precisely the same period that I have been working in the developing world. Sociologists are usually thought to be abstract, while engineers are concrete and practical. Indeed, one of the most common criticisms I have heard about sociologists—not to mention theorists--is that they are not practical. They fail to confront the real problems. They refuse to get down and dirty their hands. That could be true—but let me narrate to you exactly what has happened to our project and why I say to you now that most of the talk about closing the digital divide is just abstract and theoretical. Here I am talking about digital inclusion for the research and educational sectors of sub-Saharan Africa.

DIGITAL INCLUSION

Before you can talk about digital inclusion you must be prepared to do two things. One is to confront the legacy of "development", that is, the ways in which initiatives, programs, and projects have been imported by donors (NGOs, multilateral and bilateral aid agencies). The second is to do some work—good, old-fashioned work—where you measure distances, and get bids, and buy cables and connectors, and assess the relative merits of wireless and wired, and get up on top of rooftops and water towers and find a line of sight. Then you have got to supervise the whole thing and make sure that the work is done. I ask you, is that the work of a theoretical sociologist? I hope so. But if not, then I am happy to be a social engineer.

LOUISIANA INTERNET PROJECT

What is now the Louisiana Internet Project began in the early 1990s and focused on the production of knowledge in developing areas in the fields of agriculture and environment. After receiving funding from the Dutch Advisory Council on Scientific Research for Development Problems (RAWOO), the project collected a large amount of data in 1994 from professionals in universities, government research institutes, and NGOs in Kenya, Ghana, and the Indian state of Kerala. The focus was not the

Internet—indeed, the word was not much heard at that time—but it did generate a fairly elaborate set of data on communication and other types of linkages between individuals and organisations, both within and outside of these locations.

There are three basic findings from this 1994 study that are worth emphasising in trying to engineer the knowledge society in the developing world. First, we used data on several hundred knowledge workers in Kerala, Ghana, and Kenya to examine the character and location of their professional contacts. What we found caused us to redraw our picture of "isolated" researchers in the Third World. They are not "isolated", but rather their social ties were primarily local rather than international. Since most human populations throughout human history have had precisely this local character, it is not surprising. The scientists we studied had numerous social and professional ties but they were with others in their national research system.

The second major finding was the *negative* relationship between local and foreign professional networks. Put simply, there was a tendency for those with more ties to the developed world to have fewer local ties, and for those who had more extensive domestic contacts to have fewer linkages outside—that is, in the international arena. Whether or not that is surprising depends on one's preconception about the nature of work involving the creation of knowledge but it does have one fairly disturbing implication: that in the pre-Internet era the costs of communication (more broadly, relational involvement) meant that professional networks were a zero-sum game. It is only with great difficulty that one can have both foreign and local ties.

The third major finding was that the pattern of professional linkages to those in the developed world did not display any large differences for those who were and those who were not educated in the developed world. That is, knowledge workers educated in the U.S. and Europe had *no greater contact with Western scientists than those educated locally*. This finding seemed surprising to everyone—including me—and the meaning seemed fairly clear. Those who went abroad were quickly "reabsorbed" into the local system. Ask yourself: how often do you maintain contact with those you met in graduate school—your professors and colleagues? Our respondents in Africa and Asia did not maintain any kind of regular contact with others they had met or worked with during their postgraduate training.

CONNECTIVITY

As I was working on these issues back in the mid-1990s, the International Council for Science (ICSU) formed a Committee for Capacity

Building in Science, immediately identifying one of its three core issues as the "problem of isolation". While I am all for greater attention to knowledge processes in developing areas, I am reluctant to use that particular phraseology, given that professionals do not have any obvious deficit in the number of their contacts. Still, the promise of the Internet to me was not that "isolation" would be "cured" through connectivity. It was that the *negative relationship between internal and external linkages* might be changed. Let me reiterate this point in a slightly different way. That the Internet might make it possible for knowledge workers in developing areas to collaborate with those in the developed world is not the central issue—such collaborations have occurred for a long time. The potential is that this technology might allow an increase in all forms of communication—both domestic and international. Connectivity rapidly became one of the most important—if not the most important initiative in science and technology for development. We were not unhappy to have one of the best data sets in existence for the study of non-Internet era communication in the developing world. I only wish I could make a plausible case that I had thought of that in advance!

FOLLOW-UP STUDY

What happened next was that we began to design a follow up study of the ways that knowledge workers used the Internet, a study that would examine its effects on international collaboration, the maintenance of professional ties forged during graduate training, the activation of what sociologists call "weak ties", the distribution of types of relationships across career levels, and, of course, the degree to which these patterns vary by country and sector. We had always considered Ghana to be at the lower end of the developmental scale for our three locations, with Kenya in the middle, and Kerala—with its high level of literacy and emphasis on education—to be on the top.

During the planning phase in mid-1999 through early 2000 we found that most of the university faculties and research institutes did not have connectivity to the Internet—or had such sporadic, limited, and high cost connections that they did not function in any significant way. What we concluded was that we had to provide the Internet in order to study the Internet. These last three years have been, truly, an experience for me—because I have become a mini-development agency during the course of trying to examine what were originally just some interesting sociological questions. A half million dollars can cause you a lot of trouble when you have to give it away.

This simple fact is something that all aid agencies and NGOs know, but because they are generally not in the business of studying themselves, because they tend to have more money than I do, and because they are involved in a large portfolio of projects, they cannot spend a great deal of time reflecting on its consequences.   At the technological level, the Louisiana Internet Project is trying to do some fairly simple things. We are not doing anything as complicated as, say, building a bridge. The only bridge we are building is to the Internet. It is a matter of great excitement, of great potential, and it is worth learning how to do it. Cabling a building or configuring a wireless network for Internet access is a necessary condition for digital inclusion. Of course, it does not in itself reduce the digital divide without a wide range of other elements, elements of organisation, elements of practice and habit. But what I am emphasising here is money, resources, support. Getting practical things done takes money. I have a grant with a lot of money and I cannot spend it. That seemed strange to me at the outset— but it was a function of wanting to be assured, at some reasonable level that I cannot define, that the money would be well spent. But now it seems less strange, in light of the conditions of interactions between donors and recipients of aid that have persisted since World War II. In those terms, the connectivity initiative is simply another in a long line of Western concepts that have waxed and waned in importance, or in the more cynical view of one of my 1994 informants:

"It will be another white elephant. The donors will come in and try to establish electronic links, then leave and not support the system. It's not that important.

I continue to think it is important—especially to those professionals engaged in the business of producing knowledge. So why is it so difficult to perform these basic tasks of cabling and connectivity that are actually the preconditions of getting on with collaborative activity? What I have to say next is spoken in a code, but I think the listeners in this room are able to crack it. In many cases the recipients of aid are better at receiving money than development projects are at giving it. That fact is at the root of my statement at the beginning of this talk, that our project has yet to experience an unqualified success for institutional and relational reasons.   The institutions of development, embedded in a set of inter-organisational relations and social practices, have conspired to make it difficult to establish the connectivity that must first occur as a precondition to reducing the digital divide. "

CONCLUSION

Let me end with an illustration, a concept, and a recommendation. In the cases I am familiar with (Ghana, Kenya, and India) the actual amount of money that is really necessary to network a group of computers in offices to a properly configured server is not very much. However, when you begin to consider the logistics involved in providing the necessary amount of money for the necessary services you begin to see the problems. You cannot simply touch down in a country and go to the first store that you find. You have a limited amount of time in a country so you must work fast. You may find that payment is required before the work is done—not part now, part later. Vendors know that non-payment is very common so they do not want to do the work before the money is assured. Some vendors will flat-out refuse consideration of any job that is not backed by foreign money—if it is a Kenyan project, then they will want to be assured that the money is actually from the USAID, for instance, before going forward. On the other side of the equation, the customers desiring the connectivity do not have a very high opinion of most vendors—they often go out of business or disappear before the work is done—so there is a great reluctance to go forward unless it is with someone else's money. In eastern Kenya, together with some excellent colleagues and assistance, we cabled a building so that researchers at an agricultural institute could have a desktop connection. It cost approximately $2000, with one individual from the station providing most of the labour himself. That took more than one year—a job that our colleagues in the U.S. kept saying should be done in a couple of days. Let me give another example. At a nearby institute the cost of the materials and installation was given to me at $95,000. This is why the ideas of participation, buy-in, and cost-sharing became important. But I submit that those are not enough.

The problem has more to do with the priority of personal face-to-face relationships in developing areas, and particularly the *relationship between face-to-face contact and electronic communication*. In the example of the high bid, the key fact is that we were unable to extract the bid from the contractor until after we had left the country and email did not prove sufficient to clarify the relatively limited project requirements. We were perceived as "donors with deep pockets", in the words of one of our local collaborators. We have become enmeshed in the institution of development—but we are not developing anything. This is why we have rejected the concept of "development" and turned to "reagency". When organisational representatives from afar alight in countries bringing resources in the service of international initiatives, they take part in or bring about a reaction. We are not really interested in the representatives themselves but rather these reactions—the chains of events that occur as resources enter a matrix of personal relations in an area of scarcity. Our general thesis is that the Internet has a potential that prior initiatives did not have—because the intermittent, computer-mediated interactions can maintain relationships and projects after the donor has left, and bring about the non-immigratory friendships that can and do have a positive impact on knowledge, on social network formation, and on the success of projects.

This process of reagency leads me to the recommendation. The very best way of understanding the problems is to follow the old-fashioned advice: try this yourself. Pick a place in Africa, in Latin America, in Asia, and get it connected to the Internet. At best, you will establish a potential. At worst, you will fail, but you will understand—and you will have participated in the greatest collaboration the world has ever known.

BIOGRAPHY

Wesley Shrum has been Professor of Sociology at Louisiana State University since 1982. Since 1987 he has been Secretary of the Society for Social Studies of Science, an international and interdisciplinary association for the study of science and technology with over 1200 members worldwide. The basic aims of the society are to bring together those interested in understanding the social dimensions of science, technology, engineering, and medicine through annual meetings and publications.

Professor Shrum has been studying the social networks and communication practices of scientists and engineers since the 1970s. His first book, Organized Technology: Networks and Innovation in Technical Systems (Purdue University Press, 1985), examined the social networks of researchers involved in nuclear waste and solar photovoltaic research in the U.S. In the early 1990s his primary interest shifted to the developing world, still focused on communication and collaboration in the research process. For the past ten years he has focused on Ghana, Kenya, and the state of Kerala in south-western India. In 1994, Prof. Shrum directed comprehensive studies of the research institutions in these areas. His current studies examine the impact of the Internet on communication patterns with particular emphasis on international relations.

PROJECT PROPOSALS

The "four pillars" and e-education for all
Project proposal

Bernard Cornu
Director of La Villa Media, 22 avenue Doyen Louis Weil, 38000 Grenoble, France

bernard.cornu@lavillamedia.org

Abstract: The aim of this project is to take into account the wide set of possibilities of information and communication technologies, in order to promote and improve e-education for all, addressing each of the "four pillars" of education: learning to know, learning to do, learning to live together and learning to be. Main aims are to create new pedagogical strategies, tools and resources for e-Education for all, oriented towards the four pillars of education; to create "e-Educational services", especially for developing countries; to create tools and resources for teachers; to create international communities around e-Education strategies and to create knowledge building and exchange processes within the local community.

Key words: e-Education, e-Society, learning to be, learning to do, learning to know, learning to live together

E-EDUCATION FOR ALL

UNESCO has stated "Education for All" as an essential priority. This is a double ambition: it means ensuring access to education for everyone, especially the most disadvantaged, and it also means ensuring access to quality in education for all. Improving access to and quality of education is a major challenge for preparing the future.

Education is not only a matter of content and subject transmission. A report to UNESCO by the 'Commission for Education in the XXIst century' noted that education should include four main dimensions, four "pillars": learning to know, learning to do, learning to live together, learning to be.

Information and communication technologies (ICTs) bring new tools, new concepts, new resources and new pedagogies for teaching and learning. These new possibilities must not be reduced to just the technological tools, but must take into account all dimensions of education.

The 'e-Society', the society changed by information and communication technologies, needs a new kind of education. This new kind of education, 'e-education', musty be made accessible for all, and must be a quality education for all with respect to the four pillars of education. In particular this project is to take into account that information and communication technologies do not just bring technological changes, but more importantly imply social, global and fundamental changes. It is therefore important to relate information and communication technologies to the aims of "education for all" and to the "four pillars" of education.

PROJECT AIMS

The aim of this project is to take into account the wide set of possibilities of information and communication technologies, in order to promote and improve e-Education for all, addressing each of the "four pillars" of education.

Main aims are:
- To create new pedagogical strategies, tools and resources in order to promote e-Education for all.
- To create pedagogical strategies oriented towards the four pillars of education: learning to know, learning to do, learning to live together, learning to be.
- To create "e-Educational services", especially for developing countries.
- To create tools and resources for teachers, aiming at the development of e-Education for all:
  - Tools and resources for e-Education;

- Using e-Education for enhancing education for all;
- Using e-Education for implementing the spirit of the four pillars of education.
- To create international communities around e-Education strategies.
- To create knowledge building and exchange processes within the local community.

KEY PRINCIPLES

- Created knowledge should support the attainment of social goals, such as those stated in the UN Millennium Declaration, and should support shared understanding.
- Participation of several of the following actors: governments, UN organs and agencies, international/national/regional organisations, international professional organisations, business sector, civil society, academic institutions, and so forth.
- Projects will be directed at practical, real-world contributions to the creation of instances of e-Education, e-Health, and e-Society, especially in developing countries, and not to academic or industrial research or development. The latter may, however, be a necessary condition for the execution of the projects.
- ICT used to implement these capabilities/facilities/services must be based on the experience and finances of the sponsoring organization(s).
- ICT used to implement these capabilities/facilities/services and the businesses established to develop the local economy must recognise the lack of a strong electrical supply and of a robust communication infrastructure in many developing areas and the need to develop commerce based on dematerialized or virtual activities.
- Project execution will include a reciprocal goal of enriching the developed nations with the created knowledge of the culture, languages, and value systems of the local community.

NEEDS ADDRESSED

- Contribute to the aims of the Dakar Forum "Education for all", particularly in the field of e-Education.
- Ensure access for all to the benefits of new technologies, methods, contents, resources to improve education.

– Provide resources, methods, tools, ICT-based, in order to enhance the balance between the "four pillars" of education: learning to know, learning to do, learning to live together and learning to be.

PLAN OF ACTIVITIES

– Designing and experimenting with educational tools integrating the "four pillars" and usable in developed and developing countries.
– Producing recommendations for developing e-Education for all.
– Promoting concrete actions and experiments for "e-Education for all". For example:
  – 2 or 3 local actions in schools or universities in developing countries;
  – Distance-action through Internet about e-Education for all and the four pillars;
  – More global action at a policy-maker and decision-maker level.
– These actions should be focused on concepts such as:
  – Networking for all;
  – Enhancing collective intelligence.

EXPECTED OUTCOMES

– Concrete tools, resources, recommendations, web-resources, should be developed and produced.
– The results of the experiments, the recommendations, the tools and resources produced should be generalisable and disseminated.
– The outcomes should provide basis for "e-Educational policies". For example, following the Dakar Forum, provide an "e-Education for all Framework for Action", based on some designed and experimented actions.

CRITICAL SUCCESS FACTORS

– Real cooperation between developed and developing countries;
– Meeting the actual needs of learners and educational systems;
– Real international dimension;
– Precise contribution to the aims of the Dakar Framework for Action.

KEY MEASURES OF SUCCESS

- Local success of the first experiments;
- Possibility of generalisation and dissemination.

SCIENTIFIC RESULTS EXPECTED

- New elements in the articulation of the four pillars in Education;
- Technological, pedagogical, political, social, ethical principles and statements for developing e-Education for all.

TIMETABLE OUTLINE

1. Phase 1: Establishing principles and aims for the international project;
2. Phase 2: Describing a concrete list of possible actions to be carried out in the framework of this project;
3. Phase 3: Designing a "road-map" for each action and for the global project;
4. Phase 4: Selecting the appropriate partners and places for experiment;
5. Phase 5: Carrying out the different actions;
6. Phase 6: Gathering the results and outputs; evaluating;
7. Phase 7: Stating principles for dissemination;
8. Phase 8: From the results and outputs, design recommendations and possible actions to be stated and carried out officially as a follow up of the Dakar Framework for Action, by international bodies (UNESCO, OECD, Countries having participated in the WSIS, …).

HUMAN RESOURCES

- A coordinating board;
- Teachers and professors;
- Researchers;
- Technologists and implementers;
- Experimenters.

MATERIAL AND FINANCIAL RESOURCES

- Such a project needs resources for developing tools and methods, resources for experimenting them in some places, resources for establishing definite and permanent follow up in some places.
- Resources include: human power, computers, technology and development, resources for people to meet and cooperate.

POTENTIAL ACTORS

- Selected private companies;
- International bodies;
- European Commission;
- La Villa Media as focal point of such a project, in partnership with:
    - SATW (Swiss Academy of Technical Sciences);
    - International bodies;
    - Universities and schools;
    - Private partners;
    - Etc.

BIOGRAPHY

Prof. Bernard CORNU is director of La Villa Media (European Residence for Educational Multimedia), Grenoble, France and a professor (applied Mathematics) at IUFM (University Institute for Teacher Education) of Grenoble, France. For ten years (1990-2000) he has been the Director of the IUFM of Grenoble and, until 1994, the Chairman of the 29 IUFMs in France. He was (2000-2002) Advisor for Teacher Education at the French Ministry of Education. His scientific speciality is now the Integration of Information and Communication Technologies into Education, and its influence on the Teaching Profession and on Educational Policies.

As a member of IFIP (International Federation for Information Processing), he has been (1995-2000) the chairman of Working Group 3.1 ("Informatics Education at the Secondary Education Level"), and he is now the secretary of the IFIP TC3 (Technical Committee for Education). He is Member of the French National Commission for UNESCO, and vice-chair of the Education Committee. He is also the Vice-Chair of the Governing Board of IITE, the UNESCO Institute for Information Technologies in education, located in Moscow. He has been (1998-2002) the President of the French Commission for Mathematics Education.

E-solidarity, a means of fighting against FGM (Female Genital Mutilation)

Project proposal

Annie Corsini-Karagouni

Maasai Education Discovery Committee Member, 38, Ch. Edouard Olivet 1226 Thônex, Switzerland, tel. 0041-22-348.33.70

corsini_annie@yahoo.fr

Abstract: This project aims to contribute to the eradication of the practice of Female Genital Mutilation (FGM) throughout the Maasailand in Kenya in agreement with the World Health Organisation (WHO) policy by large-scale distribution of information to the remote Maasai villages, by creation of awareness, by proposing alternative rituals, by improvement of the social (and economic) status of women and by encouragement of Maasai families to send female children to school. e-Society means will be used in the understanding that these are not in opposition to preserving tradition and ethnic identity

Key words: e-solidarity, ethnic identity, Female Genital Mutilation, rituals, tradition, women

OBJECTIVES

The main objective is to contribute to the eradication of the practice of Female Genital Mutilation (FGM) throughout the Maasailand in Kenya. This is in agreement with the WHO's (World Health Organisation) policy, which targets complete worldwide eradication of FGM (African countries being of high priority) as of 2015.

Specific objectives are:
- Large-scale distribution of specific information to the remote Maasai village population. This essential information is centred on women's health, reproductive health, human rights, and the legal status of FGM.
- The creation of awareness among the Maasai community (Narok district had an estimated 365,750 people in the last population count of 1999) on this issue, while preserving to the utmost Maasai culture and identity.
- The proposition of alternative rituals to mark the rite-of-passage from childhood to adulthood of young girls (14 to16 years old).
- The improvement of the social (and economic) status of women by generating income based on traditional handicraft production.
- The encouragement of Maasai families to send female children to school to get an education.
- Creation of an understanding that progress and development through e-Society are not in opposition to preserving tradition and ethnic identity.

INITIAL CONDITIONS

The following initial conditions should be taken into account. The Maasai community in rural areas is a pastoral and semi-nomadic one. Living conditions are harsh; there is a chronic lack of water, sanitation, and education. ICT is an unheard-of dream in these remote areas that lack electricity, communication facilities, and roads.

Other NGOs have tried to combat FGM in the past, giving seminars in the central town and inviting people to participate. This method is partially efficient, as the most vitally concerned segment of the population never receives this information.

From a legal point of view, the Constitution and the Penal Code of Kenya prohibit FGM. And even though education is compulsory, it is not a practical reality in this pastoral semi-nomadic society.

Without any doubt, strengthening the woman's social role inside the traditional Maasai family is a key step in this project. The way to achieve this is by developing and promoting the economic activity of women.

TIMETABLE OUTLINE

The e-solidarity project includes 3 phases of action.

Phase 1

Information sessions given by health professional, volunteers and social workers in both town (Narok, the Maasailand capital in Kenya) and remote rural areas.

Access isolated villages that are off the beaten track and without electricity - areas where, to this day, e-society has no meaning.

Only personal, first-hand contact is valuable and reliable. Our visits are highly appreciated by the rural population.

Phase 2

First information is given to Masaai women about FGM, related health issues. They are being made aware of the considerable social and cultural commitment that such a decision involves. Then they organise themselves into small cooperatives that will produce traditional handicraft and Maasai beadwork such as bracelets. The MED and project leader meanwhile will identify worldwide partner web sites, which by hosting our web page would promote the anti-FGM campaign.

"Buy a Maasai bracelet and help women to say NO to FGM" will be the project identification slogan.

Phase 3

ICT becomes the connecting tool between developed and developing countries. Internet would attract and consolidate international help and propagate solidarity (by emails addressed via MED to women's groups and by selling the solidarity bracelets via the web to the international public.)

RESOURCES

To achieve the above phases of implementation, the project needs resources such as:

- Social workers and health specialists to hold open-air seminars throughout the rural villages.
- Logistic means (a 4x4 rental car to reach remote areas), educational material (the WHO provides us with this regularly), audiovisual means (TV, DVD or laptops) to show educational films and documentation.
- A MED local coordinator must keep in touch with rural women and coordinate the promotion of the Maasai women's groups' beadwork production.
- A web master, to designing the web page of the project.
- Concerted fundraising from UNESCO, UNIFEM or other human right's institutions or governmental bodies, for the large scale kick-off (over the whole Maasailand) of the implementation of the project.

PARTNERS

The Maasai Education Discovery (MED) centre in Narok will play a pivotal role in the project. Solidarity e-mails will be received through the web and centralised there. These will then be spread (read-as the women are completely illiterate) throughout the rural community.

The MED centre will gather international orders received from web ("solidarity Maasai bracelets") then will distribute these orders among the "women's groups against FGM", and finally it will collect and ship the goods to the clients.

MED's project coordinator will distribute the production benefit back to the rural women's groups, in order to allow them to launch a new, local and sustainable economic activity. He/she also will decide the start and end point of the assistance stage.

A list of international web sites, considered as potential partners, able to host a web page advertising our project has already been compiled and some agreements have been reached:

- www.eziba.com (this site has already supported the "Peace baskets project" in Rwanda)
- www.fgmnetwork.org ("education and networking projects")
- www.feminist.org (site of "Feminist majority Foundation")
- www.soroptimist.org (International Club of mutual aid)
- www.bpw.ch (Business and professional Women)
- www.cwf.ch (Carrier Women Forum)
- www.un.org/womenwatch (UN's women)
- www.cybersolidaires.org (UN's women)
- www.femmesfrancophones.free.fr
- www.whrnet.org (Women Human Rights Net)

−   www.rotary.belux.org (Rotary club) etc.

EXPECTED OUTCOMES

The following outcomes are expected:
−   Break the wall of isolation wall surrounding the rural Maasai community;
−   Strengthen the local economy through women's activities;
−   Connect the western public with social development in the remote Maasai community;
−   Develop awareness about FGM and contribute to its complete eradication;
−   Encourage the education of Maasai children (the emphasis being given to girls, victims of FGM).

CRITICAL SUCCESS FACTORS

Key factors for success are:
−   Good local co-ordination (MED's responsibility);
−   Sufficient uptake of the project by international organisations and associations having web sites;
−   Sufficient funds to implement Phase I (education);
−   Mutual trust between the local Maasai community committed to giving up the practice of FGM (and at the same time improving their social and economic conditions) and western society sensitive to and concerned by the infringement of human rights.

TIMETABLE OUTLINE

Phase 1 has already been started on a small-scale, voluntary basis.
Phase 2 is in progress. Three women's groups (totalling 80 women) against FGM have already been formed and are preparing to sell Maasai jewellery (bracelets).
Phase 3 will be soon in place as many contacts have been made in order to establish a worldwide partnership

FINANCIAL RESOURCES

The education phase 1 must find external financial resources (sponsors). The project period has therefore been projected at 3 years. The project will be implemented at Mau division first (77,686 people spread among 16,704 households). With a budget of 100'000 Euro/year, we plan to visit and sensitise 1500 households by year, employing 10 social workers or nurses. We are expecting a redaction in the prevalence of FGM and forced marriages among the target group by 50% at least within the first 6 months.

The sale of "Maasai bracelets of solidarity" is a cyclic sustainable process, having a variable time life (this is the period during which a women's group against FGM will be financially assisted) and variable extent (the area covered by our campaign will be proportional to our financial resources).

We hope to receive funds from UNESCO, UNIFEM, WHO or other international bodies.

ACKNOWLEDGEMENT

M.E.D (kenyan NGO) Maasai Education Discovery
- Mr. KOITAMET OLEKINA, P.O. Box 788, Narok, Kenya - Tel / Fax : 254-0305-23066, e-mail : Info@maasaieducation.org - Website: www.maasaieducation.org
- M.E.D, Mr. LEDAMA OLEKINA (President), 2 Park Plaza, Suite 415, Boston, MA 02116 USA, Tel : 001-617-7790619, Fax : 001-617-4263490.

BIOGRAPHY

Annie Corsini-Karagouni obtained a license in mathematics at the University of Athens in 1976. After that she received a third cycle diploma in Astronomy and Astrophysics in 1982 at the Geneva Observatory in Sauverny. Over the last 12 years she has worked for different financial institutions in Switzerland as a database administrator in IS departments. In parallel, Annie is active in various human rights and social organisations as a volunteer. She is a committee member of Business and Professional Women club (BPW) in Geneva, helps in the "Association Bleu Ciel" that is working in Rwanda, and she helps in "Sentinelles", a children rescuing Foundation in Lausanne. She is Maasai Education Discovery's director since two years and has a good contact with representatives of the Maasai community.

An indigenous approach to bridging the digital divide
Project proposal

Kenneth Deer & Ann-Kristin Håkansson

Indigenous Indigenous Media Network, Box 1170 , Kahnawake Mohawk Territory,QC J0L 1B0, Canada. Tel: +1-450-635-3050, Fax: +1-450-635-8479

easterndoor@axess.com; www.indigenousmedia.org

Abstract: Knowledge, information and communication are at the core of the emerging global Information Society. Knowledge, information and communication, however, are culturally defined concepts and expressions. Also, Information and Communication Technologies (ICTs) – the medium for disseminating and communicating knowledge and information - are cultural products of the society that has developed them. Indigenous Peoples have their own concepts of knowledge, information and communication and have developed their own forms of information communication. Therefore Indigenous Peoples need to take part in the Information Society on their own terms and on the basis of their cultural backgrounds, to be able to shape their future without risking to lose their cultures and identities.   This project aims to contribute to this process by addressing four major aspects: identification and development of culturally appropriate ICT applications; elaboration of Indigenous approaches and strategies to bridge the digital divide; design of culturally appropriate capacity-building tools; elaboration of culturally appropriate development strategies for utilisation of ICTs for poverty reduction.

Key words: culture, Digital Divide, indigenous peoples, intellectual property, traditional knowledge

INTRODUCTION

Knowledge, information and communication are at the core of the emerging global Information Society. Knowledge, information and communication, however, are culturally defined concepts and expressions. Also, *Information and Communication Technologies (ICTs)* – the medium for disseminating and communicating knowledge and information - are cultural products of the society that has developed them.

Indigenous Peoples have their own concepts of knowledge, information and communication and have developed their own forms of information communication. Therefore Indigenous Peoples need to take part in the Information Society on their own terms and on the basis of their cultural backgrounds, to be able to shape their future without risking to lose their cultures and identities.

Indigenous research projects are an important tool to assist in reaching these goals. They can support the development of Indigenous approaches, strategies and visions for the evolution and implementation of the Information Society - and thus protecting and promoting its cultural diversity.

PROJECT AIMS

The project presented here, aims to contribute to this process by addressing four major aspects:

1. The identification and development of culturally appropriate ICT applications by Indigenous Peoples;
2. The elaboration of Indigenous approaches and strategies to bridge the digital divide;
3. The design of culturally appropriate capacity-building tools for Indigenous Peoples on ICTs and their range of possible uses;
4. The elaboration of culturally appropriate development strategies for utilisation of ICTs for poverty reduction.

PLAN OF ACTIVITIES

Research is envisioned to be carried out in all cultural-geographic Indigenous regions, namely: the Arctic Region, Central America, South America, North America, the Pacific Region, Asia, Africa and Russia. Research activities will be conducted in various steps:

1. *Development of models for culturally appropriate capacity-building and information "workshops" for Indigenous Peoples.*

   These programs will be elaborated in close co-operation with Indigenous ICT experts and using Indigenous ICT experiences in this particular region. The workshops will serve to inform Indigenous Peoples on the basics of what are ICTs and how do they function as well as on the various possibilities of their application.

2. *Conducting initial surveys during these workshops*

   These workshops will be organised for each region in close co-operation with local Indigenous partners and organisations. They will also provide a platform to carry out first surveys among participants, supplemented by an analysis of workshop discussions, on the following issues:
   - Possible ICT needs of Indigenous Peoples of this regions;
   - Their views on cultural appropriateness of ICT applications;
   - Their views on culturally appropriate ways of equal participation in the Information Society on their own terms;
   - Local problems of connectivity;
   - Other obstacles towards participation in the Information Society, e.g. the question of literacy;
   - Their views on culturally appropriate strategies and approaches to overcome the digital divide in their regions.

3. *In-depth surveys in selected communities*

   The results of these initial surveys will be used as a starting point to carry out in-depth community surveys on the above mentioned issues. These surveys will be carried out in close collaboration with local Indigenous partners.

4. *Elaboration of survey summaries*

   The analysis and evaluation of these surveys will serve as a basis to identify:
   - Preliminary strategies and visions of Indigenous Peoples of the various regions towards the evolution and implementation of the Information Society in their areas;
   - Perceived challenges and potentials of the developing Information Society with regard to the survival of their cultures and identities;
   - Approaches to the establishment of an equal partnership between Indigenous and non-Indigenous actors in the Information Society;
   - Indigenous "plans of action" to bridge the digital divide in their regions on their own terms.

Research activities will be presented during a Conference on "Indigenous Peoples and the Digital Divide", which could take place as a parallel event to the Tunis part of the World Summit on the Information Society in 2005. This Conference should bring together Indigenous Peoples from the regions, who have participated in the workshops and/or the surveys, Indigenous and non-Indigenous researchers and the scientific community, Indigenous and non-Indigenous ICT experts and practitioners, Indigenous and non-Indigenous development practitioners, donors and the private sector.

This conference should be a platform to identify a plan of action to be carried out beyond 2005 to assist Indigenous Peoples to bridge the digital divide on their own terms.

BIOGRAPHY

Kenneth Deer is co-founder, editing director and WSIS focal point of the Indigenous Media Network, an international organisation of Indigenous media workers. He is the owner, publisher and editor of the weekly Mohawk community newspaper The Eastern Door, which he founded in 1992.

Since 1994 he is coordinator and often co-chairman of the Indigenous Caucus at the United Nations in Geneva. From 1971-1987 he worked in Indigenous education, partly as the director of the Kahnawake Survival School. He is founding member of the First Nations Education Council of Quebec and the National Indian Education Council (NIEC). From 1983 to 1989 he was NIEC co-chairman and took part in overseeing a $4 million study on Native education in Canada.

From 1987-1990 Kenneth Deer was Coordinator of the Mohawk Nation Office in Kahnawake, a secretariat of the People of the Longhouse (part of the Six Nations Iroquois Confederacy). From 1990 to 1992 he served as Traditional Chief of his community.

Ann-Kristin Håkansson has been assisting in producing a position paper for Indigenous representatives to Prepcom 3 and the World Summit of the Information Society. She has 20 years of experience in development co-operation with Indigenous Peoples, project management and project evaluation. Furthermore, she has carried out a number of research tasks in this context. Finally, she has organised Indigenous capacity-building programs for Indigenous Peoples from developing countries, but also served as a resource person in such training programs. Her main geographic areas of work are Latin America and Africa.

Ann-Kristin has been consultant for the European Commission, for a study on EU co-operation with Indigenous Peoples. Internationally, she participates in the United Nations Working Groups such as on Indigenous Populations and on the Draft Declaration on Rights of Indigenous Peoples.

E-well: Multi-sectoral development of rural areas
Project proposal

Antoine Geissbuhler & Ousmane Ly
Medical Informatics Service, Geneva University Hospitals, Geneva, Switzerland; Mali telemedicine network coordinator, REIMICOM, Bamako, Mali

antoine.geissbuhler@hcuge.ch; oussouly@keneya.net

Abstract: Key aspects of the "e-well"-project concern the evaluation of the impact and sustainability of integrated, multi-sectoral approaches to the development of rural areas in the least advanced countries. The development approaches include, but are note limited to, the usage of multi-purpose tele-centres and the formalisation and publication of local, collective knowledge. Aim is also a reduction of the Digital Divide, particularly obvious in rural areas of developing countries. Focus is on sustainability by fostering simultaneous development activities in multiple sectors (education, health, economy, culture) based on the assessment of local needs.

Key words: developing countries, Digital Divide, Internet access, rural areas, sustainable development, tele-medicine

INTRODUCTION

ICT has the potential to improve the quality and efficiency of cooperation and development efforts. However, the risk of ICT-enabled development projects is to further the digital divide between urban and rural areas. It is therefore crucial to involve rural areas early in these efforts, to make sure that these specific needs are addressed in national projects.

In our experience one of the key challenges in the deployment of telemedicine applications in rural Mali, as a component of the national telemedicine network (http://www.sim.hcuge.ch/telemed.html; http://www.keneya.net), is the economic sustainability of such technologies. Sustainability can be improved by fostering simultaneous development activities in multiple sectors (education, health, economy, culture), enabled by an internet-connected tele-centre, the "e-Well". This requires a significant effort, geographically-focused and involving most of the stakeholders of the community, in order to reach a significant increase in development, compatible with long-term sustainability of the process and results.

THE PROJECT

Based on an assessment of needs by the local authorities, a four-year, multi-sectoral development project has been designed for the community of Dimmbal, a community of 24 villages and 30.000 inhabitants (http://www.dimmbal.ch). Activities include: capacity building for local project coordination, implementation of a tele-centre, development of traditional medicine and its integration in the practice of the rural hospital, development of telemedicine activities with Bamako, valorisation of local history and culture, development of a museum with archaeological findings, revitalization of traditions including the Sacred Wood, and development of the local industry and forestry. The budget for this development project is estimated at € 600.000, over four years, from 2004 to 2007.

SUCCESS FACTORS

It is likely that success factors and obstacles in such projects will be educative to similar projects in other settings. The goal of the « e-well » project is to run several development projects in different rural settings in developing countries, and to evaluate, compare and share results in order to learn collectively from the various experiments. The expected outcome is a better understanding of the potential, success factors, impact and sustainability, of integrated, multi-sectoral approaches to the development of rural areas in different settings.

The "e-Well" project plans include 6 different sites with four-year development plans, and various coordination, evaluation and sharing activities between the local coordinators of each site. The project is designed to run over 7 years (2004-2010), for a total budget of € 6.500.000, under the coordination of the AGENTIS, a UNITAR agency dedicated to exploit the potential of information and communication technologies for development and social initiatives.

BIOGRAPHY

Antoine Geissbuhler is a Professor of Medical Informatics at Geneva University School of Medicine, and Director of the Division of the Medical Informatics at Geneva University Hospitals.

A Philips European Young Scientist first award laureate, he graduated from Geneva University School of Medicine in 1991 and also received his doctorate there. He was trained in internal medicine at Geneva University Hospitals. After a post-doctoral fellowship in medical informatics at the University of Pittsburgh and Vanderbilt University, he became associate professor of biomedical informatics and vice-chairman of the Division of Biomedical Informatics at Vanderbilt University Medical Center, under the mentorship of Prof. Randolph Miller and Prof. William Stead, working primarily on the development of clinical information systems and knowledge-management tools. In 1999, he returned to Geneva to head the Division of Medical Informatics in Geneva University Hospitals and School of Medicine.

His current research focuses on the development of innovative computer-based tools for improving the quality and efficiency of care processes, at the local level of the hospital, the regional level of a community healthcare informatics network, and at the global level with the development of a south-south telemedicine network in Western Africa.

Ousmane Ly is the Executive Coordinator of "Keneya Blown", the technical structure of Mali Network of Information and Medical Telecommunication (REIMICOM). Dr. Ly has a PhD in Medicine and a BSc in Biological Sciences. Currently, he is preparing his Post Medical Computing University Certificate in the University of Geneva. Mr Ly is also a member of the ATAC (African Technical Advisory Committee / United Nations Economic Commission for Africa, ECA).

Cybertroc: A barter system for the Information Society
Project proposal

Colin Harrison
Director of Strategic Innovation, IBM Global Services EMEA Strategic Outsourcing, IBM Zurich Research Laboratory, Säumerstrasse 4, 8803 Rüschlikon, Switzerland

ch@zurich.ibm.com;http://www.research.ibm.com/people/c/colinh

Abstract: Barter – or *troc* in French – is the exchange of goods or services for other goods or services. It is the oldest form of commerce and continues to have an important role in even the most highly developed societies. Not least it is highly valuable for developing social capital among a community. The Information Society brings to barter both a new meaning for "community" and new mechanisms for exchange. In particular it may be a key motivator for bridging the digital divide, but providing a tangible, understandable purpose for joining the Information Society. The goal of CyberTroc – or Internet-based barter – is to connect people with needs to other capable of satisfying these needs. The Internet is an ideal means for facilitating barter, since it can provide a transactional capability at extremely low cost and since it can bring together communities that may be highly localized or may be very scattered.

Key words: barter, community, Digital Divide, social capital

INTRODUCTION

Barter – or troc in French – is the exchange of goods or services for other goods or services. It is the oldest form of commerce and continues to have an important role in even the most highly developed societies. Not least it is highly valuable for developing social capital among a community. The Information Society brings to barter both a new meaning for "community" and new mechanisms for exchange. In particular it may be a key motivator for bridging the digital divide, but providing a tangible, understandable purpose for joining the Information Society.

The goal of CyberTroc – or Internet-based barter – is to connect people with needs to other capable of satisfying these needs. The exchange may be mutual – the two parties to a barter directly exchange goods or services of comparable value – or communal – each member of the barter community is required to maintain a personal balance of trade. The Internet is a ideal means for facilitating barter, since it can provide a transactional capability at extremely low cost and since it can bring together communities that may be highly localized or may be very scattered.

One could imagine trading many goods and services and indeed commercial bartering, for example, for the disposal of unsold lots of goods (http://barterwww.com/) or for the exchange of timeshare holiday accommodation (http://www.i-barter.com/), is already well established. However we see CyberTroc as a person-to-person activity, a core activity of the Information Society. In particular we have considered the application of CyberTroc to ride-sharing.

PROJECT EXAMPLE

An efficient mechanism for on-demand ride sharing in both urban and rural areas would have many benefits. It would promote the mobility of those who are unable to drive themselves to their destination or who are unable to afford a taxi or for whom public transportation is unavailable or ineffective. It could reduce the number and use of polluting vehicles. It could serve to develop social capital within a community and it would provide an Information Society service of direct value even to those who are otherwise uninterested. It is sufficiently simple that it could be accessible not only via a personal computer, but also via SMS or voice access. In return for the transportation, the passenger offers some good or service to the driver; since the imposition is small, the compensation need not be onerous.

A would-be traveller enters a request for a journey between a starting point and a destination and an approximate time of departure or arrival.

Others who are capable of providing transportation can view these requests, possibly as visualizations on a map, or may be identified automatically by fuzzy matching to journeys that they regularly make. The technical challenge comes from the need to make fuzzy matches between requests and offers, possibly exploiting public transportation for some segments of the journeys.

Software to implement such a CyberTroc system is available and indeed during the transportation strikes in France earlier in 2003, such a system was spontaneously created to help workers to get to their jobs (http://www.goclicktravel.com/cgi-bin/gct.pl?language=uk).

BIOGRAPHY

Colin Harrison joined IBM in San Jose, California in 1979 and has held many technical leadership positions in IBM's product businesses, in IBM's Research Division, and currently in IBM's IT services business. In 2001 he established IBM's Institute for Advanced Learning.   Following his university studies, he spent several years at CERN developing the SPS accelerator.   He then returned to EMI Central Research Laboratories in London, and lead the development of the world's first commercial MRI system. With IBM he has enjoyed a career leading from micro-magnetics to medical imaging, parallel computing, mobile networking, intelligent agents, telecommunications services, and knowledge management.

Colin Harrison studied Electrical Engineering at the Imperial College of Science and Technology and earned a PhD in Materials Science. He also studied Physics at the University of Munich.   He is a Fellow of the Institution of Electrical Engineers (UK) and a Senior Member of the Institution of Electronic and Electrical Engineers (USA). He is a Chartered Engineer (C.Eng.) and a European Engineer (Eu Ing).   He was a founder member of the Society of Magnetic Resonance in Medicine (USA). He is also an expert advisor to the Swiss Academy of Technical Sciences. He has been a visiting scientist at MIT, Harvard Medical School, and Lawrence Berkeley Laboratory.

Colin Harrison has been awarded 26 patents. He has published some 40 scientific and technical papers and talks and a successful book on Intelligent Agents. He is an invited speaker at European universities on the impact of information technology on the nature of work, business organization, and industries.

Cyber inclusion through activity
Project proposal

Jean-Marie Leclerc
Directeur général du Centre des Technologies de l'Information (CTI), Etat de Genève, Route des Acacias 82, 1227 Carouge, Case postale 149, 1211 Genève 8, Tel.: +41 22 327 79 67, Fax: + 41 22 327 48 77

jean-marie.leclerc@etat.ge.ch

Abstract: All individuals have the right to express themselves through an activity, in which one's potential can be developed and recognised. On the other hand everyone should contribute to the broad implementation and recognition of this particular right. In the cyber-inclusion project of the Centre of Information's Technologies (CTI) both technological development and inclusion of minorities, in particular handicapped people and people with marginalised lifestyles, are addressed. Three fundamental principles underpin the project process: integration, a three-dimensional approach to activities (requested, necessary and feasible) and adequate structures. The individual human being is the focus of concern.

Key words: handicapped people, inclusion, integration, marginalised lifestyle, networks

INTRODUCTION

All individuals have the right to express themselves through an activity, in which one's potential can be developed and recognised. On the other hand everyone should contribute to the broad implementation and recognition of this particular right. Accordingly, the cyber-inclusion project, of the Centre of Information's Technologies (CTI), addresses in parallel both technological development and inclusion of minorities, in particular handicapped people and marginalised lifestyles. Main aims of the project are establishment of an appropriate environment for the target population, based on their recognised potentials, and then to support the emergence of networks.

Three fundamentals principles are assisting the process. Firstly, integration is a foremost concern. Secondly, a three-dimensional approach to activities, harmonising their implementation: requested, necessary and feasible activities. Indeed, nowadays, 60% of the work related to utilisation of technologies of information and communication is not yet requested, although necessary and feasible. Thirdly, support and development must be translated into action through adequate structures, promoting cyber-inclusion as well as flexible networks. See Figure 1.

Think globally: integration
Take action locally: requested, necessary, feasible
Express yourself individually: flexible cyber networks

Figure 1. : Three-dimensional approach

We present two examples to illustrate these principles.
1. A blind employee joined our team working on information's technology. She currently realises specific tasks requesting particular skills. Her contribution considerably increased the quality of our services. Indeed, by adjusting the Internet websites of the State of Geneva to the needs of the blind citizen, this employee also helps to make clear criteria of accessibility in computer science and in technical structures.
2. CTI plays an active role in partnership with UNCSTD. Indeed, the CTI helps to create a dynamic situation in which experiences and pilot projects on e-Society in Least Developed Countries (LDC) can be capitalised. Using a strategy for action based on a vision that has emerged from experience and reflection, CTI, and particularly the Technological Observatory (TO) promises to be a strong and reliable partner for the UNCSTD.

CYBER-INCLUSION OF INDIVIDUALS INTO THE E-SOCIETY

To move on to an e-Society, involves developing specific networks, which allow the enforcement of governance favouring human sustainable development. It is fundamental not to develop these networks based on a so-called standardised representation of individuals, but rather on human factors. A maximum of individuality will prevent a new category of exclusion: cyber exclusion.

The definition of exclusion is logically connected to the one of inclusion. Indeed, criteria of inclusion define de facto the reasons for exclusion. But this is not sufficient to deal with the issue: powerful majorities, by lobby or by consensus, also play a role in the delimitation of this vague concept. However, besides economical and social criteria, moral and ethics also enlarge the number of constraints in the system.

Nevertheless, every society, by its integrating power, logically also produces exclusion. Unfortunately education, history, believes and moral code reproduce these exclusions. Modern society does not escape this perverse cycle: barbarous exclusions are to be encountered on account of race, health abilities, sexual orientation, nationalism, employment, etc.

If no action is undertaken, our modern life, based on technologies and information science, will have dividing power in two different directions: numeric exclusion and economical exclusion.

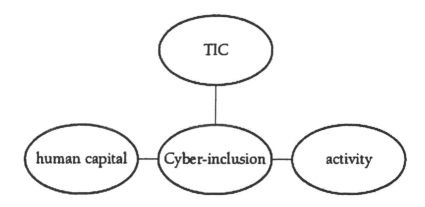

Fig. 2: articulation of the cyber-inclusion

The current challenge is not to suddenly eradicate exclusion and injustices from the world, but rather to benefit from the enlargement of world wide communication and networks to bring certain patterns to awareness. 'e-Society', as a sub-ensemble of common society, is therefore submitted to the same rules: injustice will go on. The main concern is that this new type of society will not add a new type of exclusion, but rather that it will allow a new type of inclusion for those excluded by other criteria of other sub-ensembles. This task is divided into three axes of action (see Figure 2.).

TECHNOLOGICAL CYBER-INCLUSION (TIC)

Technological devices (computers, software, Internet networks, etc.) and legal structures are the gate keepers of the access to e-society for individuals and more general actors.

The integrating power of TIC, however, falls short in case of handicaps. Technologies extend our potential in communication, calculation, programming, simulation, task execution and networked teamwork through simple adaptations on the logistical level. A Braille line connected to the computer is an example of how simple the criteria for inclusion may be enlarged to blind population.

CYBER-INCLUSION THROUGH KNOWLEDGE

Quoting the Organisation for Economic Co-operation and Development (OCDE), human capital is defined as: "knowledge, qualifications, skills and other qualities possessed by an individual". Therefore, the more individuals can enlarge their human capital, the broader and the richer will e-Society be. Contrary to some tendencies in our society to keep back information and keep competency at a low level, the dynamics of e-society turns this around. The most important skills that need universal distribution are information management and human capital management. Flexibility in dealing with information and abstract material is one of the fundamental competences needed to participate in e-Society's activities. Qualification is a practical issue in the economic sphere and/or civil society.

THE THREE-DIMENSIONAL APPROACH
APPLIED TO ACTIVITY

Following the principles previously introduced, three types of activities are to be encountered:

1. *Necessary activities*: activities needed by social and economic spheres in order to be reliable.
2. *Requested activities*: outcomes of demands expressed by potential clients, ready to entrust others with a mandate.
3. *Feasible activities*: operational human capital, able to realise requested activities. These human profiles allow efficient performance to realise entrusted mandates.

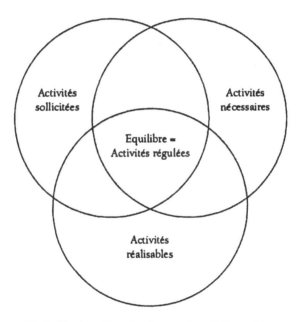

Fig 3 : The three-dimensional approach applied to activity

The zone located at the intersection of the three circles corresponds to the best balance between the various activities. They consequently represent 'regulated activities'.

HUMAN BEING AS MAIN CONCERN

Social cohesion as well as social capital can result form the cyber-inclusion approach, but only if networks are developed and empowered. Indeed, they appear to be the best environment to let human capital develop through activities and exchange of knowledge and skills.

Networks, boosting human capital, are the answer to subdivided and centralised power related problems. The classical pyramidal (hierarchic) organisation slows down initiative and reduces autonomy. On the contrary, network organisation fosters transversal relationships, and especially vertical and transversal circulation of information. Contained information is a loss of potential. The e-approach prevents subsidiary circulation of information, which divides the monitoring process from the activity process. The raw material of e-Society can therefore be defined as knowledge and skills. Thus, a human and reticular structure delimits the framework of action for e-activities. Inside this framework, autonomy is highly recommended and interdependency is preserved by interconnection and exchange of information.

Human concerns also integrate economical constraints, as trade represents one major key to sustainable development. Thus, extension of human capital, through networks and using a cyber-inclusion approach, should not be perceived as a 'charitable attitude', but rather as the only coldly rational approach to remain competitive in e-business and e-trade.

BIOGRAPHY

Jean-Marie Leclerc is director of the Centre of Information's Technologies (CTI) in the State of Geneva (Etat de Genève) since March 2001. Main milestones in his career are:
- PhD, University of Compiègne, Paris, France;
- Thesis dedicated to the conception of information's systems in decentralised and pluridisciplinary environments;
- International mandates, particularly in the field of development of information science, related to health systems;
- Lecturer at the Neuchâtel's University and various Schools of engineer in Telecommunication and Management, Switzerland;
- Development of multilateral exchanges among public administrations, especially in the region of Neuchâtel (inter-administration project), Switzerland.

Exploring dilemmas: Ethics, social values and e-Society
Project proposal

Deryn Watson

Professor of Information Technologies and Education, Department of Education and Professional Studies, King's College London, Franklin-Wilkins Building, Waterloo Road London SE1 9NN, United Kingdom

deryn.watson@kcl.ac.uk

Abstract: All agents for change come with a baggage - and there is now an uneasy balance between technological imperatives and opportunities on the one hand, and social and ethical values on the other. It is up to all stakeholders in society to accept the responsibility of ensuring that this e-world is informed by a range of perspectives and societal values. In particular we need to find a balance between technology shaping social events and vice-versa. Any knowledge society should be giving its citizens a sophisticated digital literacy that enables them to understand the e-world. A knowledge society would include a critical understanding of both the dilemmas that societies and individuals now face in this new e-world, and the different ways in which they might seek to address them. This ‹ Dilemmas › project is rooted in the epistemologies of the social sciences and aims to help expose the multiplicity of perspectives and routes towards understanding issues associated with societal values in an e-world.

Key words: digital literacy, ethics, information age, responsibility, social values, vulnerabilities

DILEMMAS FOR KNOWLEDGE SOCIETIES

Society has been following a heady route into the ‹ information age ›, but has only recently become aware of its limitations and dangers as well as its value and potential. All agents for change come with a baggage - and there is now an uneasy balance between technological imperatives and opportunities on the one hand, and social and ethical values on the other. One of the strongest drivers of the information age has been the technological world itself coupled with business and commercial interests. It is up to all stakeholders in society to accept the responsibility of ensuring that this e-world is informed by a range of perspectives and societal values. In particular we need to find a balance between technology shaping social events and vice-versa.

One example of the problem of balance has been the identification of digital divides. But not all divides are between the developed and developing worlds, the advantaged or disadvantaged. Because the complex reality of societal values are driven by the different histories, politics, cultures and traditions in which our societies are grounded. In some ways, all societies and their members are currently made vulnerable by this e-world - threatened by concerns such as privacy, crime, security, surveillance, legality, control, discrimination and rights. Engineering a knowledge society that supports human development demands that such vulnerabilities and related ethical dilemmas are addressed. And yet there will be differing means of addressing these vulnerabilities depending on how they are experienced in local contexts.

Any knowledge society should be giving its citizens a sophisticated digital literacy that enables them to understand the e-world. Those responsible for education must ensure that all citizens, young and old, understand the structures, concepts and organising norms of the e-world, in order to use it critically and be knowledgeable enough to make informed decisions. But these structures, concepts and organising norms are not simple. Indeed even digitalisation itself is an extreme form of abstraction. It may be insufficient for citizens to simply have the practical skills of using information technologies. Without a sophisticated understanding of the concepts and values embedded in these technologies, they will be unable to participate in deliberations about the dilemmas they bring, and be subject to pressures exerted on them by interested parties.

A knowledge society would include a critical understanding of both the dilemmas that societies and individuals now face in this new e-world, and the different ways in which they might seek to address them. A knowledge society would not be one which simply imposed one group's values on others, but one which understood and tolerated differing solutions for differing societies. And finally, a knowledge society would find ways to harness the e-world to suit its aims and values.

A "DILEMMAS" PROJECT

This ‹ Dilemmas › project is rooted in the epistemologies of the social sciences - here learning is often contexted in problems with no single correct answer, or definite route for arriving at a solution. Thus the project aims to help expose the multiplicity of perspectives and routes towards understanding issues associated with societal values in an e-world. The process of learning, discussion and collaboration generated, is as valuable as ‹ finding an answer ›. This is particularly important when social values are both culturally determined and contested.

The project seeks to use a range of contexts, settings and cases to identify paradigm dilemmas posed by information technologies for social values, explore how different groups are tackling these dilemmas, and reflect on the differing responses offered by different communities. The framework of relevant themes to be addressed in the project will include :
– How far does technology create or limit choices ?
– Can individual rights and informed choice be protected within frames of social control ?
– How can technology both empower and disempower ?
– Can there be an acceptable balance between censorship and individual rights ?
– How to balance the advantages and threat of digital surveillance ?
– How to protect the vulnerable from predators ?

The project would select a range of contrasting international contexts and settings, for example an educational institution (school, university), a community (commune or village, city authority) large organisation (professional, business, charity) and government department (regional, national). The selection would be on the basis of a range of differences regarding scale, nature of an organisation and cultural environment, rather than perceptions of advantaged or disadvantaged. Thus the contrasts between cases would reflect histories, politics, cultures and societal values.

The purpose of the project would be that project participants would identify dilemmas in their own terms. The way dilemmas are discussed, responses to the dilemmas, and the constraints participants felt they were under, would be used as a basis for exploring comparable processes for others. This analysis would clarify and elaborate the paradigm dilemmas in this field. This would form a basis for developing concrete strategies and tools for others to use.

This analysis of dilemmas, and the different ways societies deal with them would have two outcomes. Firstly it would bring into the open the very real dilemmas that the e-world creates for human society, and the processes which groups and individuals use to seek compromises or solutions to address these dilemmas. Secondly it would provide rich and real examples of how societies, and individuals within them, interact; providing these grounded insights is itself an important part of social science education and understanding.

Indeed, knowing how to recognise and explore these dilemmas broadens our definition of e-literacy.

BIOGRAPHY

Deryn Watson, Professor of Information Technologies and Education, and Head of the Department of Education and Professional Studies at King's College London, studied Geography at Cambridge. After teaching in London schools, she became Humanities Director of the Computers in the Curriculum project (Chelsea College), developing computer assisted learning materials in the humanities, their potential for interactive learning, and exploring models of software development. She then researched the impact of IT on children's achievements, factors influencing the adoption of IT use in teaching (her doctorate, London University), and ICT and change in teacher education. Current research interests include issues which influence the use of ICT in education, professional development and change, and the social and ethical issues of the 'Information Age'.

Deryn Watson is a member of the Education Committee (TC3) of the International Federation of Information Processing (IFIP), and Chair of the working group of ICT and Informatics in secondary education (WG 3.1). She has served on committees of a many international conferences, recently Social, Ethical and Cognitive Issues in Informatics and ICT (Dortmund 2002). She has given keynote addresses at conferences in Canada, Australia, Taiwan, and the Caribbean, is Editor-in-Chief of the international journal, Education and Information Technologies (Kluwer Academic), and an invited expert of the Swiss Academy of Technical Sciences.

Network-blended education of tomorrow
Project proposal

Tom J. van Weert & David Wood
Chair ICT and Higher Education, Hogeschool van Utrecht/Cetis, P.O. Box 85029, 3508 AA Utrecht, The Netherlands, Tel: +31 30 258 6296, Fax: + 31 30 258 6292

School of Psychology, University of Nottingham, University Park, Nottingham, NG7 2RD. United Kindom. Tel: +44 115 951 5302 Fax: +44 115 951 5324

t.vweert@cetis.hvu.nl; http://www.cetis.hvu.nl; djw@psychology.nottingham.ac.uk

Abstract: The introduction of ICTs is causing considerable tension between systems like those of education, health and democracy and what is happening outside those systems. This tension has to be resolved. Therefore educational goals need to be redefined. In redefining the goals of education changes need to be directly related to competencies and how the acquisition of these competencies can be integrated into the curriculum and the evaluation system. This project aims are: creation of new pedagogical strategies; creation of international communities around these new strategies in higher education and upper secondary education; creation of new principles and methods; creation of knowledge building and exchange processes.

Key words: higher education, international communities, knowledge building, knowledge exchange, pedagogical strategies, redefinition of education, upper-secondary education

INTRODUCTION

The introduction of ICTs is causing considerable tension between systems, like those of education, health and democracy, and what is happening outside those systems. Therefore one of the major challenges of the Knowledge Society is institutional change. Institutions are the building blocks of society and, as such, the future of society depends on our ability to adapt and/or develop institutions that structure and give sense to our lives. Much important learning takes place outside the educational systems. This situation fundamentally challenges the pertinence of institutional based learning. Educational institutions have considerable difficulties reconsidering their relationship with learning taking place outside their own limits.

The tension described has to be resolved. Therefore educational goals need to be redefined. In redefining the goals of education changes need to be directly related to competencies and how the acquisition of these competencies can be integrated into the curriculum and the evaluation system. This area represents one of the major axes of work for the future: defining goals, devising new structures, identifying competences, elaborating ways of developing these competencies, creating suitable forms of evaluation.

PROJECT AIMS

– Creation of new pedagogical strategies, especially in developed countries, but also sustainable in developing countries;
– Creation of international communities around these new strategies;
– Primary target groups: higher education and upper secondary education;
– Creation of new principles and methods for the exploitation by the members of an educational community, irrespective of age, of the skills, resources, facilities, and so forth;
– Creation of knowledge building and exchange processes within the local community.

Answers to questions are sometimes available elsewhere, but are not always accessible. The reference here is to the barriers between particular areas of activity, when it comes to the flow of information. This is particularly the case between research and teaching practice. The dynamics of the research context are often such that there is little incentive to communicate results to anyone other than fellow researchers. This situation is unacceptably wasteful. We need a more "ecological" approach to knowledge and its development. There are often fundamental differences in perspective between researchers and those working in the field, like teachers requiring a considerable effort to establish exchange of knowledge and experience between these actors. A possible answer might lie in some form of "co-learning".

KEY PRINCIPLES

- Participation of several of the following actors: governments, UN organs and agencies, international/national/regional organisations, international professional organisations, business sector, civil society, academic institutions, and so forth.
- Projects will be directed at practical, real-world contributions to the creation of instances of e-Education, e-Health, and e-Society, especially in developing countries, and not to academic or industrial research or development. The latter may, however, be required for the execution of the projects.
- The "local community" must include balanced representation of all segments of the community, including advocates for the old and wise, the young adults, and the next generation.

NEEDS ADDRESSED

- Improving the fitness of graduates for 21C society;
- Lifelong learning.

PROJECT APPROACH

- Networking of knowledge, access to knowledge, human beings, and objects;
- Blending of ICT with other educational resources by integrating the abstractions of ICT with the principles of pedagogy.

The first step to shaping the modern world is developing a shared vision based on a clear idea of what is happening. The idea of developing a "vision" is the first step towards launching appropriate activities in the so-called "Information Society", in particular as far as the digital divide is concerned. Note that there is not one "digital divide", but many. For example

North/South,   East/West,   poor/rich,   men/women,   young/old, town/countryside, trained/untrained, ... The aim of having a "shared" vision is above all to promote the transparent discussion of values and goals in a world where much of the driving motivation behind action goes unchallenged and un-discussed.

EXPECTED OUTCOMES

- New model for the role of the teacher;
- New model for the organization and operation of education;
- New basis for the application of ICT in education.

CRITICAL SUCCESS FACTORS

- Access to a number of real institutions willing to undertake this experiment;
- Ability to engage both ICT and teachers to work as a community in these institutions;
- Willingness of educational authorities to recognize the students' achievement for graduation;
- Willingness of educational authorities and teachers to recognize the value of the new approach.

KEY MEASURES OF SUCCESS

- Recognition by employers that these students are better prepared for entering the workforce;
- Ability of the new model to be reproduced spontaneously;
- Achieving better than the Hawthorne effect.

SCIENTIFIC RESULTS EXPECTED

The scientific result of the project will be a validated theoretical model for the integration of ICT in networked education.

The project should be based on the idea of action research: Integrating development, content, research and use. Following on from the conclusion of the European eWatch project, it is argued that all activities in the education should be organised around "research communities" involving software developers, content developers, teachers, supporting staff and research workers. From the research perspective this would be called "action research".

TIMETABLE OUTLINE

Year 1	Establishing working arrangements with a number of institutions.
Year 2	In vitro, small-scale experiments on networked-education and deployment of ICT and other infrastructure; evaluation
Year 3	Small-scale application and development of syllabus for a small number of grades; staff development; creation of international communities.
Year 4	Initial deployment and continued development of syllabus; staff development; international community involvement.
Year 5	Continued deployment and development of syllabus; international community involvement.
Year 6	Continued deployment and development of syllabus; initial assessment of overall results; international community involvement.
Year 7	Continued deployment and development of syllabus; final assessment of overall results; international community involvement.
Year 8	Continued deployment and expansion; international community involvement. . . . . and so forth.

HUMAN RESOURCES

- 5 researchers in pedagogy (5 academic institutions);
- 5 researchers in ICT and other media technologies (5 academic institutions);
- 5 implementers (5 academic institutions);
- 4 technology operators per educational institution (10 institutions);
- 10 teachers (half time) per institution (10 institutions)

A core group of universities be set up who agree to evolve such a global vision and implement it in institutional strategy and practice. To give body to these ideas and to translate them into concrete actions a North-South twinning of schools using ICTs may be developed with a view to developing a relationship that seeks to avoid "neo-colonialism". That is to say, the setting up of a two-way exchange of ideas and knowledge that not only respects diversity but considers it as an immense source of richness.

Another action concerns mobilising universities to implement the propositions given here. Universities were chosen because they represent a key step in providing skills and knowledge for professional activities and are relatively close to the professional world. One could argue that the whole education system should be concerned, but universities have greater freedom than schools or colleges in determining their policies and obtaining additional funding and as such are more able to implement the necessary changes. The major question is going to be to what extent existing academic culture and the related ways of working can be modified through a process designed to elaborate a shared vision.

MATERIAL RESOURCES

- Network facilities per year ($100k/academic-year);
- ICT and media infrastructure per institution ($100k/institution);
- Face-to-face community meetings (2/year) ($1M/year).

ACKNOWLEDGEMENT

In the text use has been made of a vision document prepared by Alan McCluskey, Bern, after a SATW-meeting in Gwatt, Switzerland

BIOGRAPHY

Tom J. van Weert holds the chair "ICT and Higher Education" of the Hogeschool van Utrecht, University of Professional Education and Applied Science, The Netherlands. His main research interest is in Lifelong Learning of professionals and its implementation in Higher Education. Tom has been managing director of Cetis, Expert Centre for ICT-based Innovations in Higher Education of the same university. Before this he was director of the

School of Informatics (Computing Science) of the Faculty of Mathematics and Informatics of the University of Nijmegen, The Netherlands.

Tom has studied applied mathematics and computing science starting his career in teacher education. He has been chair of the International Federation for Information Processing (IFIP) Working Groups on Secondary Education and Higher Education. Currently he is vice-chair of IFIP Technical Committee 3 on Education with special responsibility for TC3 Working Groups. He also is invited expert of the Swiss Academy of Technical Sciences (SATW).

David Wood is Director of ESRC, Centre for Research in Development, Instruction and Training, University of Nottingham. Principal research interests centre on the nature of instruction and learning with particular reference to developmental disabilities such as childhood deafness and learning difficulties. Current work also includes research which brings together expertise in instructional theory, Artificial Intelligence and Human-Computer Interaction to develop new architectures for Intelligent Tutoring Systems and Learning Environments.

ICTs for education and development in rural communities
Project proposal

Sithabile Magwizi

*Global Teenager Project Coordinator / Facilitator, World Vision International Zimbabwe, 59
Joseph Road, Mount Pleasant, Harare, Zimbabwe. Tel : +263 4 301715 / 301709 / 301172;
Fax : +263 4 301330, Mobile : +263 91 331369*

stabile_urenje@wvi.org

Abstract: Proficiency in ICT skills is now regarded as important as basic reading and
               writing skills. In order to achieve computer literacy among the entire
               population, ICTs should be incorporated into the formal education system of a
               country. It must be noted however, that this process does not begin and end
               with putting computers in schools. Maximum use and benefits can only be
               derived through corresponding changes in approach to teacher training,
               curriculum development and administration. Overall, the support from
               teachers and the community is necessary to ensure the success of strategic
               initiatives. This can only be achieved by involving input from representatives
               of all stakeholder groups, for the development and implementation of the
               necessary policies and strategies. To gauge the extent to which this goal
               achieves transformational development, which is sustainable, practical
               measures will have to be put in place.

Key words: curriculum development, rural areas, teacher training

KEY ISSUES

For development and implementation of the necessary policies and strategies for ICT development in education, the key stakeholders in the sector include:
– Students;
– Teachers;
– Parents;
– Education Institutions
  – Early childhood;
  – Elementary and secondary school;
  – Postsecondary;
  – Special education (For people with disabilities);
– Libraries and resource centers;
– Policy planners and administrators;
– Student loan and financial aid administrators;
– Career guidance and training providers;
– Employers;
– Authors, publishers and local book stores;
– Designers and suppliers of educational software;
– Vendors and service providers.

The following questions should be used as a guideline for identifying the key issues that need to be addressed in relation to the development of ICT in education:
– In what ways can ICT be used to enhance education delivery and learning in the classroom?
– How will teachers be trained, and how should they be motivated to use the technology?
– How can ICTs be employed to alleviate the high student to teacher ratio, and to extend the reach and capacity of the education system?
– What method should be employed to obtain the most equitable distribution of the limited ICT resources among the various rural schools?
– How to strike a balance between investing in ICTs in schools and spending on more immediate and basic educational needs (Can we justify the return on investment)?
– What methods can be employed to obtain and maintain the supporting physical and technical infrastructure required for ICT implementation?
– What steps should be taken to ensure a closer alignment between ICT education and the world of work?

- In cases where investments have already been made in ICTs in schools why is the resource still being utilized?

STRATEGIC OPPORTUNITIES

"ICT has huge potential to engage pupils in ways that will help to realize their individual potential, whilst also offering teachers new opportunities to develop their professional skills in the classroom" – Estelle Morris

The concept looks at combining access to the Internet and networked computers to produce an innovative breakthrough child focused solution that is geared at achieving sustainable transformational development in the communities we work with. Effective use of ICT can enrich and enhance all aspects of schooling; teaching and learning, management and administration, and pupil's achievement.

Improve educational delivery and learning

- Computers can be used in classrooms to enhance learning by improving the quality of education in the classroom. Computer aided instructions gives opportunity for fast learners to flourish and slow learners to catch up through facilitation of self-paced learning. The Internet can be used in the classroom as a research tool to broaden horizons of both teachers and students.
- Equip school leavers with the ICT, information and learning skills needed for employability and lifelong learning and enable them to engage in a technological society.
- Support innovations in schools improving the effectiveness of schools and teachers.

Facilitate distance education

Distance education can be used to strengthen educational capacity and also to provide equitable access to limited resources especially in remote areas. Thus ICTs can be used to introduce or enhance distance education, by facilitating online courses and e-learning for both teacher and student.

Distance education can provide a very useful supplement for the shortfalls in higher education and professional training institutions.

Strengthen administration and decision making capacity

ICT can be used for effective monitoring and management of the education system. (e.g. tracking and analysis of student performance, student record keeping, tracking of fee payment, etc.)

Establishment of community resource center

Computers in schools are usually used during school hours. There is possibility of extending computer use to provide:
1. Adult education and computer literacy classes on evenings, weekends, and during school vacation.
2. Other professionals (e.g. business people, medical personnel, etc.) in the community, e-mail and Internet services to communicate with family and friends all around the country or world.
3. Typing, faxing and printing services. This would benefit individuals in the community as a whole, and would reap a quick return on investment by charging a nominal fee.

Extension of research facilities

Creates the opportunity to link schools, libraries, resource centers and research facilities locally and internationally.

CHALLENGES AND THREATS

Effective implementation of policies and ideas necessitates identifying and forecasting potential challenges:
- Lack of ICT Expertise among policy makers;
- High opportunity cost of technology;
- Limited budget allocation for maintenance;
- Shortage of teachers with ICT skills;
- Possibility of widening achievement gap between schools.

CRITICAL SUCCESS FACTORS

- Collaboration between major stakeholder groups is necessary to ensure successful implementation of these strategies.
- Physical infrastructure must be upgraded in schools in preparation for ICT introduction.
- Basic literacy is the foundation to computer literacy, so emphasis should not be shifted from basic literacy in favor of computer literacy.
- Regional and international collaboration is important to help learn from the success and mistakes of other countries.
- Curricula and teaching methods must be modified to incorporate the use of technology.
- Teacher involvement in planning and change is crucial for ensuring acceptance and support.

FRAMEWORK FOR IMPLEMENTATION

Unless ICT development is incorporated into the development plans of the school they risk misplaced investment and wasted opportunities for learners. Unless we are clear about what we want to achieve we will not be able to plan or achieve:
- Framework for implementation;
- Budget must be available and approved;
- Physical infrastructure upgraded in terms of power,space and security;
- Site inspection and collaboration with the school and teachers nominated to run the project;
- Network, hardware, software installations and Internet connection;
- Training;
- Commission;
- Networking (Try to find partners who are willing to add value to the project).

IMPACT OF TECHNOLOGY ON SOCIETY

Human societies, behaviours, systems of knowledge, and technologies are all interwoven in continual change. To describe the impact of some particular technology on society is a reductionist approach that risk missing the impact of society on the technology. Even when technologies that have been developed in centres of industry are exported to centres of continuity without any adaptive change to their design, they might be used in ways their designers had never imagined. (This is parodied in the opening scenes of the South African movie, *The Gods Must Be Crazy*. A soft drink bottle tossed from an airplane comes to serve an unmanageable variety of competing uses for the nomadic community that lives where it happens to fall.) These impact areas could be:

- Physical;
- Psycho-intellectual;
- Professional (skills for work);
- Economic;
- Cultural;
- Social;
- Environmental (e.g. changes in the use of physical space);
- Political.

JUST IN TIME SUPPORT FOR EDUCATIONAL LEADERSHIP
Project proposal

Niki Davis
Director of Iowa State University Center of Technology in Learning and Teaching, Iowa State University, Ames, Iowa 50011-3192, USA., Also: Professor of Information and Communication Technology, School of Maths, Science and Technology, Institute of Education, University of London, 20 Bedford Way, London WC1 0AL, UK.

nedavis@iastate.edu, n.davis@ioe.ac.uk; http://www.ctlt.iastate.edu, http://www.ioe.ac.uk/mst

Abstract: In the World Summit on the Information Society we recognise four worlds: the technologically advanced first world; countries in transition; third world countries with the biggest challenges; a fourth world within all those regions and nations: the poor side of the digital divide. A comprehensive synthesis of research into Information and Communication Technologies (ICT) in education produced to inform the World Summit on the Information Society provides a cautious appraisal of the value of ICT for education (Davis & Carlsen, this volume). This synthesis of research into ICT in education also questions the transfer of research from the first world into more challenging circumstances. The question is: "How can capacity be created for mentoring on ICT in education?" This project aims to create the infrastructure to facilitate networking between graduate students and leaders of their home regions. Through such ICT-enabled networks students studying abroad who gains skill and knowledge with ICT in education could provide the "just in time" ICT mentoring that a leader in their home region needs.

Key words: capacity building, educational leaders, graduate students, mentoring, networking, research, transfer

INTRODUCTION

In December 2003 as we meet in Geneva for the World Summit on the Information Society we recognise four worlds:
1. The technologically advanced first world with the highest GDP;
2. Countries in transition;
3. The third world with the biggest challenges;
4. A fourth world within all those regions and nations of the disenfranchised – also recognised as the poor side of the digital divide.

A comprehensive synthesis of research into Information and Communication Technologies (ICT) in education, produced to inform the World Summit on the Information Society, provides a cautious appraisal of the value of ICT for education (Davis & Carlsen; this volume). This synthesis of research into ICT in education also questions the transfer of research from the first world into more challenging circumstances. Essentially the synthesis recognises that the educational gains related to the application of ICT require thoughtful systemic planning and application of ICT. Positive results from the application of ICT in education appear to require the following conditions: teachers with deep understanding of pedagogical application of ICT in education, with good access for themselves and their students to educationally relevant ICT both in school and outside school. Without this pervasive application of ICT the potential gains from investment of ICT for educational benefits are fragile and may even be negative (thus working against education). However, we have relatively little research and guidance for good practice beyond the first world and research in other contexts needs to be promoted. Research and development in low income regions is also likely to benefit the fourth world, reducing the digital divide within high income countries. There is therefore an urgent need for critically aware and culturally sensitive approach to planning ICT in education outside the first world accompanied by extensive support and mentoring where possible.

We also recognise that the information society depends on ICT and that engagement for all populations requires skills with and knowledge of ICT. Relatively few leaders outside the first world have this experience, although many do have an appreciation of the concepts and the means to support their knowledge. They have many additional challenges in the leadership of their regions, cultures and society. Essentially they need 'just in time' support to make sense of ICT in education as they plan for it. But how is that to occur? One theme of evidence from research into ICT in education is secure; access to education can be extended through ICT. Distance education can be enhanced through ICT. That access could also provide leaders and policy makers with appropriate knowledge and mentoring. Then the question becomes: How can capacity be created for mentoring on ICT in education?

Many of the cleverest people from challenged regions use education to improve their circumstances. They work to gain an advanced education often with financial support from government agencies; unfortunately this often leads to a "brain drain." Clever young people who study abroad often do not return to their home region despite their original intent. Often the return may be too challenging and these "emigrants" do not have the social networks to enable them to fit back into their societies where their education may be put to work effectively. The problems of reintegrating after study abroad are well documented. ICT could help these clever postgraduate students studying abroad to be of service to their home regions during their studies. The knowledge gained through reciprocal-mentoring of leaders and the related social networks created during such a process could also help these clever people return home and thus decrease the brain drain.

This project aims to create the infrastructure to facilitate networking between graduate students and leaders of their home regions. Through such ICT-enabled networks students studying abroad who gains skill and knowledge with ICT in education could provide the "just in time" ICT mentoring that a leader in their home region needs. The successful strategy of reciprocal mentoring developed by ISU Centre for Technology in Learning and Teaching to develop ICT in teacher education in the USA (Chaung, Thompson & Schmidt 2003) would be further developed to prepare exemplary ICT-teacher education for contexts in other parts of the world.

PROJECT AIMS

– Just in time support for leaders of ICT in education;
– Reduction of brain drain into first world;
– Development and research of ICT in education to develop good practice for the second and third world, plus the digital divide (the fourth world).

KEY PRINCIPLES

– Leaders of teacher education and of curriculum development need to plan with knowledge of ICT in education (see for example UNESCO, 2002, planning guide on *ICT in Teacher Education*).

- Reciprocal mentoring of leaders by students studying abroad who have gained skills and knowledge of ICT in education could provide just in time mentoring with support from their advisors.
- Growth of ICT in education will be best developed using ecological principles building capacity with social systems that start from those existing today.
- Good practice and research in education is context and culture sensitive because ICT must be applied to serve educational aims.

NEEDS ADDRESSES

- Planning for ICT in education in second, third and fourth worlds;
- Curriculum and teacher development in relation to ICT appropriate to regions and cultures;
- Decrease of the 'brain drain' with study abroad.

PLAN OF ACTIVITIES

Development of a detailed action plan and its evaluation. Development is to be phased, starting with pilot, moving to medium phase, and finally scaling up. Within each phase:
- Recruitment and detailed strategy development;
- Set up learning pairs of leader and graduate student, ICT infrastructure and material (adaptation of existing materials already offered, see below);
- Reciprocal mentoring takes place;
- Participant research supported by experts alongside;
- Evaluation and setting up progression plan;
- Evaluation of this phase;
- Dissemination of results alongside these activities;
- Progression to next phase.

EXPECTED OUTCOMES

- Informed leadership working together with appropriate support from expertise in first world;
- Plans and strategies for ICT in education appropriate to specific regions and cultures;

– Better planning and retention of the diversity of education and other cultural aspects in the Information Society;
– Redesign of educational research in ICT in education to include many more dimensions related to first, second, third and fourth worlds and their many diverse cultures.

CRITICAL SUCCESS FACTORS

Recruitment of participants for each phase. For example, first phase would recruit approximately five of the following:
– Leaders in second and/or third world;
– Doctoral students from that region and culture;
– Educational organisations that support those doctoral student;
– Researchers to support accompanying research (may be the same as above).

Also:
– Project management, leadership and evaluation (formative and summative). This could occur through Iowa State University Center for Technology in Learning and Teaching and/or the Institute of Education, University of London.
– Funding to support this project and its growth, possibly including specific student grants.
– Acknowledgement of the potential value and support from agencies such as UNESCO.
– Multinational support, such as Cisco and/or other ICT industry.
– Political will to promote the proposed collaboration across these generations.
– Support from research and educational communities, possibly facilitated by early dissemination at International Federation of Information Processing at WCCE'05 in South Africa.

SCIENTIFIC RESULTS

– A comprehensive review of research that becomes applicable to more regions and cultures in the world;
– Systemic educational research world wide, for the first time;
– Networks that link policy makers with researchers in ICT in education.

RESOURCES

New resources

A comprehensive grant to establish the pilot phase described above, possibly from agencies such as the World Bank.

Scholarship grants for doctoral students specifically related to this project.

Existing resources offered to the project:

– Institutional commitment from Iowa State University CTLT and its doctoral program by Niki Davis (Director and Professor) including potential commitment from students from Turkey, Malaysia, China, Ukraine, Jamaica, Colombia etc. Also in Institute of Education, London potential commitment from students from Chile and Brazil.
– Modules in distance learning via the web for Virtual University of Finland through Helsinki University, Finland [http://www.edu.helsinki.fi/iqform/].
– Modules for administrators' education from Schoeny (2003) and the University of Virginia Center for Technology and Teacher Education.
– Potential links with ISTE standards for administrators [http://www.iste.org/nets] and the international committee of the Society of Information Technology for Teacher Education [http://www.aace.org/site].
– Human networking support from UNESCO Institute for Information Technologies in Education, Moscow. [http://www.iite.ru].

REFERENCES

Chuang H., A. Thompson & D. Schmidt (2003) Faculty technology mentoring programs: Major trends in the literature. *Journal of Computing in Teacher Education*, 19, p. 101-106.

Davis N.E. & R. Carlsen (this volume) *A comprehensive synthesis of research into Information and Communication Technology in education*. In T. van Weert (ed.) Education and the knowledge society. Information technology supporting human development. Kluwer, Amsterdam.

Shoeny Z. (2003) *Effective technology integration garners educational rewards* http://e--tiger.org and five courses for educational administrators on http://www.alter1.org

UNESCO (2002) ICT in teacher education. A planning guide. UNESCO: Paris, France. [http://unesdoc.unesco.org/images/0012/001295129533e.pdf]

University of Helsinki (2003) IQ FORM Research Group two courses to prepare educators for the Virtual University of Finland [http://www.edu.helsinki.fi/iqform/]

BIOGRAPHY

Niki Davis is Director of Iowa State University Center of Technology in Learning where she leads the graduate program in Curriculum and Instructional Technology that is well known for its emphasis in teacher education. She also holds a Chair in ICT in Education at the Institute of Education, University of London, where she is a member of the London Knowledge Lab. Before this she held a chair in Educational Telematics in the University of Exeter in the UK where she set up the Telematics Centre.

Niki has researched information technologies extensively, particularly in teacher education and in flexible and distance learning. She is currently the President of the international Society of Information Technology in Teacher Education and Chair of the International Federation of Information Processing Technical Committee 3 on Education's Working Group on Research. She is also an invited expert of UNESCO on ICT teacher education.

DECLARATIONS

Unesco – IFIP Youth declaration
IFIP World Computer Congress 2002

IFIP World Computer Congress 2002
International Federation for Information Processing, Hofstrasse 3, A 2361 Laxenburg, Austria, Tel: +43 2236 73616, Fax: +43 2236 736169

ifip@ifip.or.at; www.ifip.or.at

Abstract: Unesco's commitment to enhancing the participation of all in the global information society and IFIP's role in analysing and shaping future development of Information and Communication Technologies (ICTs) have inspired participants in the IFIP World Computer Congress 2002 "Information Technology for our Times: ideas, research and application in an inclusive world" (25 to 29 August 2002, Montreal, Canada) to develop declaration, having examined the theme of "Youth and Information and Communication Technologies, Policies and Challenges in the Information Age". ICTs have substantial impact on today's world and are central to bolstering the emerging global knowledge information society. Young people are at the forefront of technological innovation and development. On the other hand there is continued deterioration of the status of youth worldwide (particularly of young women and youth with disabilities), who are among the most vulnerable and affected by difficult social and economic conditions. This highlights the importance of sensitising governmental authorities, national and international institutions, the private sector and the civil society to the necessity to include the development of information and communication technology infrastructures and the ICT skills for young people as a high priority in their national ICT policies and respective agendas, as well as to take proactive measures in order to encourage the formulation of policies and regulatory frameworks determining the future of the information society.

Key words: access, culture, economics, education, ethics, information, human interaction, Lifelong Learning, open learning, social cohesion

INTRODUCTION

We, participants in the IFIP World Computer Congress 2002 "Information Technology for our Times: ideas, research and application in an inclusive world", held from 25 to 29 August 2002, in Montreal, Canada, having examined the theme of "Youth and Information and Communication Technologies, Policies and Challenges in the Information Age", have adopted the following declaration.

CONTEXT

- Taking into account UNESCO's commitment to enhancing the participation of all in the global information society, and IFIP's role in analysing and shaping future development of Information and Communication Technologies (ICTs);
- Noting the substantial impact of ICTs in today's world and convinced that ICTs are central to bolstering the emerging global knowledge information society;
- Considering that, beyond their role in economic development, ICTs can contribute significantly to building new partnerships and interactions and spreading innovative lifelong learning opportunities;
- Further considering that the universal access to information and human interaction, by means of ICTs is essential for achieving goals of social cohesion, and economic and cultural empowerment;
- Recognising the need to promote digital inclusion in an environment preserving cultural diversity and heritage and promoting the respect for democratic values, human rights and tolerance;
- Realising that some young people are at the forefront of technological innovation and development;
- Being concerned about the continued deterioration of the status of youth worldwide (particularly of young women and youth with disabilities), who are among the most vulnerable and affected by difficult social and economic conditions, and who face, among others, a growing rate of functional illiteracy and unemployment, poverty and conflicts, epidemic diseases, substance abuse and HIV/AIDS pandemic, etc.;
- Highlighting the importance to sensitise governmental authorities, national and international institutions, the private sector and the civil society about the necessity to include the development of information and communication technology infrastructures and the ICT skills for young people as a high priority in their national ICT policies and respective agendas, as well as to take proactive measures in order to encourage the

formulation of policies and regulatory frameworks determining the future of the information society;
– Affirming our commitment to contribute to ensuring a youth oriented digital inclusion specifically in the fields of education, science, culture and communication.

RECOMMENDATIONS

We strongly recommend the following measures for empowering youth in the Information Age:
– Promote global access to information and knowledge sources of young people as a prerequisite to their competent social choice, behaviour and participation; disseminate information about issues having a practical impact on the every day life of young people.
– Improve access to education and train young people in ICT skills enabling them to enter empowered into the Information and Knowledge Society; improve network access at affordable cost, especially in underdeveloped urban, rural and remote areas, and expand information infrastructure for human development through the establishment of vocational schools at a community level, the creation of internet access points, distance learning and community multimedia centres, etc.
– Provide for the equitable expansion of the information society by promoting ethics in cyberspace through the involvement of young people in the elaboration of guidelines for the activities of information and content producers, users and service providers.
– Strengthen the capacity to generate knowledge and indigenous production of freely accessible contents, while using local languages and thus expanding the existing information accumulated in the public domain.
– Facilitate the production and dissemination of high quality free and open source software for education and training as well as scientific and cultural purposes.
– Enhance the co-ordination of youth information related programmes and resource mobilising efforts of governments, specialised agencies, intergovernmental and nongovernmental organisations, and invite international and national institutions and the private sector to design and implement specific funding schemes and programmes such as fellowships, competitions and contests, that would help improving the meaningful access of young people to ICTs especially in the developing countries.

– Promote through the use of ICTs, specific measures and modules for enabling disabled and handicapped youth to participate more actively in society.
– Cultivate creativity, open life-long learning opportunities for young people and promote their access to careers dealing with ICTs.
– Support the efforts of youth to foster a culture of peace, tolerance, sustainable development and quality of life by using global information and communication means.
– Commit ourselves to strive according to the spirit and letter of this declaration for the implementation of the above recommendations.

Montreal, August 29, 2002

Vilnius declaration

IFIP World Information Technology Forum (WITFOR)
International Federation for Information Processing, Hofstrasse 3, A 2361 Laxenburg, Austria, Tel: +43 2236 73616, Fax: +43 2236 736169

ifip@ifip.or.at; www.ifip.or.at

Abstract: Participants from 68 countries at The First World Information Technology Forum (WITFOR), organised by IFIP under the auspices of UNESCO and hosted by the Government of Lithuania, gathered in Vilnius, Lithuania, 27-29 August 2003. They addressed the following major goals: Bridging the digital divide; Ensuring the freedom of expression; Reducing poverty through the use of education and Information and Communications Technology (ICT); Facilitating social integration; Respecting linguistic and cultural diversity; Fostering the creation of public domains; Supporting communities in fighting illiteracy; Encouraging e-governance and e-democracy initiatives; Improving the quality of life; and Protecting the local and global environment. The Forum conducted its work through 8 commissions: Preparing the ground for ICT; Building the infrastructure; Economic opportunities; Empowerment and participation; Health; Education; Environment; and Social and ethical aspects of the Information Society.

Key words: cultural diversity, Digital Divide, e-democracy, e-governance, environment, freedom of expression, illiteracy, linguistic diversity, poverty, public domain, quality of life, social integration

INTRODUCTION

Participants from 68 countries at The First World Information Technology Forum (WITFOR), organised by IFIP under the auspices of UNESCO and hosted by the Government of Lithuania, gathered in Vilnius, Lithuania, 27-29 August 2003, have addressed through the Forum the following major goals:
- *Bridging* the digital divide between rich and poor in the world; urban and rural societies; men and women; and different generations;
- *Ensuring* the freedom of expression enshrined in Article 19 of the universal declaration of human rights and other such instruments;
- *Reducing* poverty through the use of education and Information and Communications Technology (ICT);
- *Facilitating* the social integration of excluded segments of societies;
- *Respecting* linguistic and cultural diversity;
- *Fostering* the creation of public domains with full respect of intellectual property rights;
- *Supporting* communities in fighting illiteracy;
- *Encouraging* e-governance and e-democracy initiatives;
- *Improving* the quality of life through effective health service systems;
- *Protecting* the local and global environment for future generations.

CONTEXT

We as participants are:
- *Aware* of the complexity facing national governments in developing reliable and affordable ICT;
- *Further aware* of the importance and need of safe and secure ICTs as the foundation of global, regional and local Information and Communication services, supporting governments, organizations, enterprises and individuals;
- *Convinced* that governments need to build upon ICT- related achievements and independently evaluate existing pilot projects from the perspective of beneficiaries;
- *Subscribing* to the importance of safeguarding the economic, social, environmental and cultural rights of all peoples, with special attention to the protection of traditional societies and indigenous people;
- *Believing* in the equitable and ethical sharing of the benefits of ICT and the minimisation of any negative impacts;

- *Fully* accepting the realities facing often demanding partners, especially in the sector of economic investment required to set up the physical infrastructure;
- *Conscious* that most of the discussions on the future of the information society is being driven by technology push more than by citizens' needs

DECLARATION

We call upon national governments, civil societies and other partners to commit themselves to the implementation of the above stated objectives and to translate their commitment to the development of ICT through the creation of a favourable environment for partnership and economic investment. We resolve to work closely with all the above-mentioned partners and commit ourselves to the following strategic actions:
- *Inviting* national governments to give priorities to national socio-economic development plans for the creation of ICT infrastructures through:
  - International co-operation among central governments and through international development agencies;
  - The establishment of public and private partnerships as the cornerstone of the deployment of ICT at the local and national levels;
  - Facilitating investments in the physical infrastructure by international and regional financial institutions;
  - Supporting the development of new ICT tools and contributing to international programmes for ICT advancement;
  - Ensuring affordable and equitable accessibility to ICT between urban and rural communities and between men and women, taking into consideration the existing generation gap.

- *Urging* national governments to guarantee the application of the principles of freedom of expression and privacy through appropriate legislation that will:
  - Implement these principles as they apply to traditional media, also to the Internet, and satellite broadcast;
  - Promote public access to data and information of public interest which is held by governments, private organizations or companies.

- *Ensuring* a continuous process of education on the rights of citizens as a fundamental element of poverty alleviation by:

- Facilitating affordable universal access to the Internet and encouraging networking and dialogue between the diverse communities of interest.

- *Facilitating* knowledge and information sharing (especially as it affects the rights of the poor and the excluded) and facilitating their progressive integration into the fabric of cities, towns and societies to reduce existing social tensions and conflicts.

- *Encouraging* international cooperation for the provision of safe and secure information and communication networks and systems.

- *Supporting* the development and adoption of free and Open Source solutions wherever more affordable and /or suitable than proprietary solutions.

- *Promoting* a harmonious society within the cultural diversity of countries, convinced that national languages must never be seen as an obstacle to access to ICT.

- *Facilitating* an environment and a physical and legal infrastructure for the establishment of public domains where:
  - Universal access to content is guaranteed as an essential part of the freedom of expression with due respect to legislation governing the rights to intellectual property.

- *Empowering* all communities, especially grass roots communities, through systematic programmes aimed at developing literacy, including ICT literacy, which progressively involve community members in cooperative actions.

- *Encouraging* the use of new ICT tools, especially with regard to the new development paradigm in e-governance and e-democracy:
  - Giving due regard to social and ethical aspects and the special needs of different groups in society;
  - Empowering them to benefit from the digital revolution.

- *Promoting* the use of ICTs to address the basic needs of communities, particularly by creating a modern social health system that would improve their quality of life with special emphasis on:

- Targeting major health problems in developing countries notably HIV/AIDS, TB, Malaria and mother and child health care, through effective health management information systems;
- Optimizing the use of free and open source software, models and component specifications in future health information systems;
- Intensifying training and education in local adaptation, maintenance and use of health related information systems.

- *Improving* the use and application of ICTs in projects aimed at protecting the local and global environment for future generations, and in developing systems for monitoring potentially environment-threatening process and systems that will ensure a continued healthy environment.

We, the participants, representing national governments, business communities, NGOs, IGOs, academia and international organisations invite all partners to translate the above strategic actions into implementable action plans. We call upon all national, regional and international financial institutions to be involved in the implementation of these action plans by investing in the necessary development of ICTs at local, regional and national levels.

BACKGROUND

The Forum conducted its work through 8 commissions:
1. Commission 1 Preparing the ground for ICT;
2. Commission 2 Building the infrastructure;
3. Commission 3 Economic opportunities;
4. Commission 4 Empowerment and participation;
5. Commission 5 Health;
6. Commission 6 Education;
7. Commission 7 Environment;
8. Commission 8 Social and ethical aspects of the information society.

Below follows a summary of their findings.

Preparing the ground for ICT

Availability and use of ICTs across a spectrum of public and business domains is rightly highlighted as a crucial area of government action for development. To date, however, there has been a depressingly low rate of success in such efforts, largely due to an overly technocratic approach to the problem. ICT policies for the diffusion of technology must be made in the context of development priorities and be accompanied by actions to create socio-economic conditions that enable local communities to appropriate ICTs for the improvement of their lives. To prepare for ICT and successful participation in the Information Society, there needs to be sustained government action in understanding and responding to social demand, reshaping national education, building indigenous science/technology/engineering capabilities, and effecting economic and social reform.

Understanding and responding to social demand includes creating awareness regarding opportunities offered by ICT for socio-economic change, as well as the effort involved in such change; use of ICT in government organisations; interventions to empower socially excluded groups.

Reshaping national education includes tackling illiteracy; development of computer and information skills; professional engineering and management capabilities; development of critical abilities.

Building indigenous science/technology/engineering capabilities includes cultivating a socio-economic context of innovation; tertiary education curricula for engineering and management; local R&D balancing export opportunities and local needs, international R&D.

Effecting economic and social reform includes measures for appropriate economic liberalization and the development of an effective market socio-economic regime; support of entrepreneurial activities, social policies to alleviate the destructive effects of socio-economic innovation; legal frameworks for the information economy; political mechanisms enabling citizens to participate in local and global socio-economic change and negotiate their preferred life conditions.

Building the infrastructure

There are three themes for ICT development:
1. Connectivity to the communications infrastructure;
2. Capacity to make effective use of ICTs;
3. Content of information and knowledge stored on and transmitted by the ICT networks.

Infrastructure should not be considered as only equipment, but should be understood as: Technology, Services, Human resources, Legal and regulatory frameworks and Economy. Some recommended actions for building a sustainable ICT infrastructure are:

- Adherence to International standards should be encouraged and when appropriate enforced.
- Operation and maintenance should be considered from the very beginning of any ICT project to ensure project success and sustainability. *"Do not invest in what you cannot maintain".*
- Promotion of universal Internet access should be supported to minimize the risk of widening the digital divide.
- Implementation of new ICT services should be encouraged.
- On-going education at three different levels: executive awareness level,
- user skills level and professional technical level.
- Governments should formulate a long-term vision, set policies, regulate, protect the users and control the quality of services.
- Regulatory authorities should be in charge of a comprehensive regulatory framework to enforce efficiency, competition, transparency and universal access.

Economic opportunities

Developed and developing countries are divided by a multitude of economic, social, cultural and other issues, but arguably the most significant divide at present is digital. Combined with the globalization of trade, new digital technologies are presenting economic opportunities and creating wealth at a rate that threatens to increase the divide. The level of developing countries involvement and use of ICT slowly improves but many countries are unable to make real progress in economic development without assistance.

A comprehensive set of national policies and strategies prepared by WITFOR´s Economic Opportunity Commission provides clear paths to bridge the digital divide. Timing is a key factor. If developing countries are delayed in their digital development then the rapid growth of the developed world may leave them impossibly far behind.

Empowerment and participation

With ICT governments is able to improve the quality and expand the reach and accessibility of the services they offer to their citizens. In that way good governance is the goal and e-Government is the way. Building up modern governance in the information society the approach has to be citizen-centred, cooperative, seamless and polycentric.

There are four postulates that have to be stated for achieving good practice (according to the 2003 eEurope Awards Conference):
- e-Government is the key to good governance in the information society;
- e-Government is impossible without having a vision;
- e-Government is not just about technology but about a change in culture;
- e-Government is not just about service delivery but about a way of life.

Up to now the low take up of public e-services has been a serious problem. To improve this situation, some conditions have to be met:
- Services must become less bureaucratic and citizens have to get economic and individual benefits.
- The needs of specific target groups must be addressed.
- A multi-channel access mix is necessary with a diversity of contact points: home and mobile, kiosk, citizen office and multifunctional service shops.
- A single-window access must be provided for all services regardless of government level and agency.
- Knowledge enhancement must be interwoven into service processes.
- Special promotions must concentrate on individual groups of addressees: rural and traditionally underserved communities; the younger.
- Democratic decision-making has to stress citizen participation. For this the interaction between individuals and organizations has to be sustained by electronic services.

Health

IT strategies in health care should target the major health problems in developing countries, such as HIV/AIDS, TB, Malaria and mother and child health. Developing countries should therefore prioritise *Health Management Information Systems*, using multiple sources of aggregated and anonymised data from different related sectors in society, aiming at strengthening health management and primary health care delivery including a basic hospital structure. *Integration* within and between health care establishments requires the specification of data sets and terminology to be consistent.

Future health information systems should optimally use *free* and *Open Source* Software, models and component specifications characterised by: scalability and flexibility through a component-based architecture enabling free combination of relevant services allowing for incremental development; portability separating logical and technological specifications; and a fine-grained architecture to reduce complexity.

Sustainable systems must be based on: training and institutional development enabling local adaptation, maintenance and use; leadership of health professionals and other domain experts in systems development; and must focus on the local use of information for action.

Education

Nature of education
The nature of education is to improve a person's relation to theworld. The organization, methods, structures and objectives of education should be brought into alignment with the knowledge society.

Lifelong learning
ICT can allow education to be spread around many communities, and promote lifelong learning and capacity building for the whole community.

E-inclusion
ICT should be used to reduce the education inequalities. Women, unemployed and disadvantaged people (refugees, disabled, etc.) should receive special attention in this process. National contents must be developed in locallanguages.

Computer literacy
At all levels of education, computer literacy and ICT competence for the knowledge society should be achieved, adapted to local conditions.

Teacher education
Any educational system reform should start with teacher in-service and pre-service education. Teachers should be encouraged to acquire and use ICT equipment and skills.

Environment

Need for identification and capacity building
- Identification of special needs of Least Developed Countries (LDCs) and indigenous people, safeguarding of their local culture.
- Capacity building efforts leading to self-sufficiency in all aspects of training and supporting programs and technologies in ICT for the environment.

Public policy and access
- Development and implementation of equitable strategies for ICT, including access to affordable systems and connectivity.
- Decreasing non-tariff barriers on ICT and the environment.
- Promotion and establishment of policy alternatives that provide for equitable growth and development.
- Promotion and establishment of policies that promote the links between
- environmental knowledge, data quality and human health, environmental degradation, natural disasters, climate change, food security, water supply and quality, and other related issues.
- Improvement of public accessibility and understanding of ICT and environmental issues. This includes the roles of public data access and data sharing between organizations, regions and countries.

Monitoring and regulatory issues
- Identification, development and promotion of the role of ICT in environmental monitoring processes and the regulation of environmental issues. This includes local or regional monitoring for regulatory processes and monitoring as part of regional or global networks (for example, global climate change monitoring networks).
- Establishment of networks for monitoring that provide equitable access to facilities, equipment, training and communications so that there is both data sharing and technology transfer.

Social and ethical aspects of the Information Society

Most of the discussions on the future of the information society suggest that it is being determined by technical feasibility and driven by technology push more than by users' and customers' needs. Little attention is paid to social impact and ethics – except in the field of education and culture.

Among the social and ethical concerns we strongly suggest a focus on professional ethics; access to content and technology for all; education, literacy and public awareness; multilingualism, cultural concerns; influence of globalization; regulation, self-regulation, governance and democratic participation; intellectual property rights; specific digital policies such as eHealth, eWork, eGovernment, etc; privacy; protection of human and civil rights; protection of the individual against surveillance; development of the quality of life and well-being; combating social exclusion; computer crime, cyber-attacks and security; employment and participative design at work; risk and vulnerabilities.

It is recommended to establish national or regional social and ethical committees to assess these issues and develop social and ethical benchmarks, to ensure that the balance between technical and social aspects is maintained. If we are to enter an information and knowledge society we must remain critical and human.

Carthage declaration
On the Digital Divide

World Congress on the Digital Divide
World Federation of Engineering Organisations (WFEO), Maison de l'UNESCO, 1, rue Miollis, F-75732 Paris Cedex 15, France, Tel: +33 1 45 68 48 46, Tel: +33 1 45 68 48 47, Fax: +33 1 45 68 48 65 (new)

l.bialy@unesco.org ; http://www.unesco.org/wfeo/

Abstract: Representatives of the engineering and technology community, gathered in Tunis October 14- 16, 2003 as part of the preparatory process for the World Summit on the Information Society, agreed on the following principles. Information and Communication Technologies (ICTs) offer enormous potential to generate and distribute wealth, but it is essential that the digital gap between the "info rich" and the "info poor" be closed. The World Summit on the Information Society is seen as a forum for debating the issues and advancing viable solutions by engaging the public/private sector in partnerships. The proven ability of innovation and research-development to find solutions to problems generated by the new Information Society is emphasised, but the development of technology must not be guided solely by profit

Key words: development, Digital Divide, "information poor", "information rich", innovation, research

INTRODUCTION

Representatives of the engineering and technology community gathered in Tunis October 14- 16, 2003 as part of the preparatory process for the World Summit on the Information Society by the World Federation of Engineering Organizations. There was active participation of representatives from:
– The Tunisian, Swiss and Senegalese governments;
– The International Telecommunications Union;
– UNESCO, the World Bank, the United Nations Economic Commissions for Africa and western Asia, The International Satellite Organization, the Technical Park of Trieste, the World Innovation Foundation.

PRINCIPLES

The representatives have agreed upon the following principles:
– *Information and Communication Technologies (ICTs) offer enormous potential to generate and distribute wealth* and contribute to the United Nations Millennium Development Goals and World Summit for Sustainable Development Plan of Action.
– *It is essential that the digital gap between the «info rich » and the « info poor » be closed.* The digital gap contributes to the widening of the economic gap and aggravates exclusion and marginalization.
– *There is strong concern regarding the growing disparities of network access.* The disparities extend beyond the North-South and exist within countries, between generations, and different social classes.
– *The World Summit on the Information Society has a convincing potential* as a forum for debating the issues and advancing viable solutions by engaging the public/private sector in partnerships.
– *The proven ability of innovation and research-development* to find solutions to problems generated by the new Information Society is emphasized.
– *The development of technology must not be guided solely by profit*; science, engineering and technology must serve the needs of people.
– *The determination to actively work as partners* to reduce the Digital Divide is affirmed. Engineers are on the front lines, their fundamental role is to adapt science for the benefit of people, in particular the poor.
– *A vision on the Information Society is affirmed*: a society which is open and inclusive, promotes the diffusion of knowledge and facilitates the sharing of information. A society that values the development of human beings above all else, one that respects cultural and linguistic diversity.

– *A call is put to governments* to agree to commit resources to create a vehicle for financing low cost high speed network access that enables the sharing of knowledge and technologies meeting basic human needs for water, food, energy and health. It is strongly suggested to use the Solidarity Fund created by the United Nations in order to meet above mentioned goals.

VIENNA DECLARATION
On Wireless Local Area Networks (WLANs)

Conference on Security in Wireless Local Area Networks
European Academy of Sciences and Arts (EASA), Mönchsberg 2, A-5020 Salzburg, Tel: +43/662/ 84 13 45, Fax: +43/662/ 84 13 43

office@european-academy.at;presidential.office@european-academy.at; http://www.european-academy.at/de/index.html

Abstract: In October 2003 the European Academy, together with Österreichische Nationalbank, held a conference on Security in Wireless Local Area Networks (WLANs). During this conference, speakers, members of the audience and organisers agreed upon recommendations regarding WLAN usage and implementation. In light of the rapidly increasing acceptance and use of WLANs, the European Academy of Sciences and Arts seeks to ensure that security risks arising from these open networks and the lack of security standards do not increase accordingly. The areas concerned are: technological security, security management, legal security and privacy awareness.

Key words: legal security, open networks, privacy, security risks, technological security, wireless

INTRODUCTION

In October 2003 the European Academy, together with Österreichische Nationalbank, held a conference on Security in Wireless Local Area Networks, chaired by Dr. Hellmuth Broda, EASA/Sun Microsystems. During this conference, speakers, members of the audience and organisers agreed upon the following recommendations regarding WLAN usage and implementation. In light of the rapidly increasing acceptance and use of Wireless Local Area Networks (WLANs), the European Academy of Sciences and Arts seeks to ensure that security risks arising from these open networks and the lack of security standards do not increase accordingly (for details on individual presentations see the « Downloads » link at http://www.europeanacademy.at).

TECHNOLOGICAL SECURITY

a) New security standards will have to be defined by the industry (see standards set by personal area networks such as Bluetooth).
b) Before a wireless network is implemented, its necessity should be questioned in the light of today's security issues. Upon its implementation, accurate event log files are vital to security and for tracing security gaps.
c) For the user, seamless connectivity irrespective of the connection protocol (e.g. WLAN and UMTS) is desirable.
d) Ease of use: researchers and developers are called upon to make their products as easily accessible and understandable for their users as possible, without sacrificing security.
e) Users should be encouraged to acquire a certain degree of technical competence and knowledge of the underlying physics (e.g. optimal positioning of antennae) in order to be able to make informed security decisions. A European Security Certificate for users would be worth considering.

SECURITY MANAGEMENT

a) Security is achieved through products: security is a continuous management process. The entire communication/data path needs to be made secure; the security of a system is determined by its weakest link.
b) Trust management builds upon security management. Both need to be seen as ongoing, continuous processes, and approached in a methodical,

inclusive way. Transparent policies on data protection and handling will add to users' trust. Audits and quality seals can play an important role in this process.

c) The awareness of both users and providers regarding security concerns must be heightened, and a sustained framework for quality created.

d) Management concepts must be inclusive, taking behavioural, legal, social, organisational, technical and economic aspects into account.

LEGAL SECURITY

a) A comprehensive analysis of the legal aspects and implications of WiFi based on requirements (e.g. Basel II) regarding IT infrastructure is urged.

b) Legislative reforms should not be made without careful consideration. Should they be necessary at all, new regulations should be created in a minimalist fashion.

c) Laws concerning network and data security must be clarified and interpreted, and should converge on a European level.

PRIVACY AND AWARENESS

a) Management of authentication and confidentiality should be seen as key factors for overall security. In order to more effectively protect privacy and confidentiality, further research with the goal of heightening security must be encouraged and supported by the private and public sector (e.g. by the European Commission).

b) Education in security know-how and awareness should be incorporated into school and university curricula.

c) Widespread acceptance of WLANs depends in part on the cost structure for usage, which today is often prohibitive.

d) Ethical behaviour should be promoted and integrated into everyday use. Legal stopgaps and regulations are not sufficient.

COMPREHENSIVE, INTERDISCIPLINARY APPROACH

The conference agreed that only a comprehensive approach and ongoing interdisciplinary co-operation will enable us to build trust and confidence in wireless network technology.

Index